Fragile Science

Also by Robin Baker

Sperm Wars
Baby Wars
Sex in the Future

ROBIN BAKER

Fragile Science

The Reality Behind the Headlines

MACMILLAN

First published 2001 by Macmillan
an imprint of Macmillan Publishers Ltd
25 Eccleston Place, London SW1W 9NF
Basingstoke and Oxford
Associated companies throughout the world
www.macmillan.com

ISBN 0 333 90102 9

1 3 5 7 9 8 6 4 2

A CIP catalogue record for this book is available from
the British Library.

Typeset by SetSystems Ltd, Saffron Walden, Essex
Printed and bound in Great Britain by
Mackays of Chatham plc, Chatham, Kent

To Reg and Daisy . . . and travels just ended.

To Romany Sarah . . . and travels just begun.

Contents

Acknowledgements

Give or take the odd president of the United States, there are no names of individuals in this book – and very few names of universities, institutes and companies. This is a matter of policy, not oversight. In some ways this is a critical book and in order to tread the middle ground over some of the major scientific hypes of our time, it has been necessary to point out shortcomings in some very widely accepted research. This means finding fault with some cherished ideas and some cherished pieces of work. Sometimes the implication is that the science may be wrong, sometimes that there are forces at work other than the search for scientific truths. These might seem good reasons for naming names – but they are not. Science correctly proceeds by making mistakes, recognizing them and then hopefully eventually correcting them. And some of the forces that override the scientific process are very powerful, difficult for any scientist to oppose or ignore. My interest is in the subject matter – how it affects our lives, and how much or how little of it we should actually believe – not in witch-hunts. But although nobody is named, this book could not have been written without the decades of painstaking – if sometimes misdirected – research and writings of multitudes of scientists. I sincerely thank them all – and hope that one day they will each get the recognition they deserve.

Throughout my own thirty years as a research biologist, I was acutely aware of how fragile science could be, both in my own field and in others. Shortcomings in the scientific process itself and the nepotism that masquerades as the peer review of both manuscripts and grant applications conspire to push research programmes in favoured and fashionable directions irrespective of validity. And when commercial and media interests also become involved, the scientific process can become brittle indeed. A book on the frailties of high-profile science always seemed very appealing. It wasn't, though, an easy book to write, not least in the choice of subjects to cover. I could have made life easy and less contentious for myself by concentrating on areas that had been unambiguously proven to be flawed: for example, the search for treatments for morning sickness that culminated in the thalidomide tragedy, or the decades of theorizing about stomach ulcers that were turned on their head when the culprit was found to be a simple bacterium. But rather than produce a matter of record, I wanted instead to write a book that focused on topical subjects and explore the chances that these, too, might be flawed.

As I considered the way in which each topic should be handled, my partner Elizabeth Oram was my strongest critic. She was the first hurdle each chapter had to clear before it was shown to the outside world. Catherine Whitaker at Macmillan then applied her keen editorial eye and saw immediately what was working and what was not. And throughout the process, my agent Laura Susijn was always at the end of the phone line to advise, cajole, criticize and encourage – usually gently. I thank all three for their combined wise advice and encouragement.

Biological Times

More than ever before, the fruits of biological research – which I take to include medical research – are becoming a staple part of our daily diet. Almost every news bulletin – and certainly every daily newspaper – now has its statutory item on health or the environment. GM foods, destruction of the rainforests, global warming, skin cancer, heart disease, mental health, and so on vie with each other to stir our emotions, generating fear or passion. And as each new issue arises, we turn to biologists for information and advice. Our questions are simple. For example, 'Is it safe to eat beef?' or, 'How dangerous is sunbathing?' Sadly, though, biology is a complex science and there are no simple answers.

Partly for this reason, the biologists themselves – tongue-tied by the precise jargon of their subject – rarely answer us. Instead, an eager media intervenes, armies of journalists who on our behalf scour scientific journals and research labs around the world. They are looking for stirring stories, written in science-speak but ripe for translation into simple prose. Little is plainly presented, of course. It is fashionable to blame the media for the public face of science – but the public clearly yearns to be alternately alarmed and reassured and to be given quasi-evangelical causes. To be newsworthy, biology has to be crusading, frightening or reassuring. If the media are tempted to titivate biology, it is only because most people would otherwise find it boring. Given this, can we be certain that scientific truths are being fairly presented by the media? Can we even be certain that there are scientific truths in the first place? Perhaps the science itself is so unreliable – so fragile – that it does not merit our emotional energy, a simple case of much ado about nothing. The science itself could be faulty, nobody yet knowing the real truth. Or there could be pressures so great that neither scientists nor media are putting truth above all else.

Consider the simple pressure of wanting to be popular. Biologists are well aware that some views will be more appealing to the media and the public than others. Suppose research showed that saving the giant panda from extinction was neither biologically nor economically justifiable. Who would be the more popular: someone who campaigned to save the panda's habitat come what may or someone who stayed true to their science and campaigned to develop the area commercially? Then there is the even greater pressure of political correctness. Suppose a biologist discovered, beyond reasonable doubt, that rapist genes are present in all men. Rapists, therefore, are merely men unlucky enough to find themselves in a situation in which they cannot control themselves. De facto, all rapists should be allowed the claim of diminished responsibility. Who would dare publish such a finding?

And biologists make mistakes. Again, this is partly because they are human, impatient for fame and funding when a small pilot experiment seems to produce tantalizingly earth-shattering results. But it is also because biology is a science, and in science mistakes are not really mistakes. Current truths are nothing of the sort; they are merely the best contemporary insight into a situation. And no matter how logical such insights might seem, they exist only for as long as it takes new research to prove them wrong. As a science, biology need have little concern for these mistakes; the mistakes of the past are simply stepping stones to the truths of the future. But to everybody else – all those who depend on biological advice to organize their lives and safeguard their health and future – such mistakes do matter.

So how good is the science behind the media hypes of our time? The aim of this book is to strip away political correctness and popularity and examine as objectively and clearly as possible the scientific basis of some major contemporary concerns. In the process, a number of uncomfortable questions have to be asked. What if the public face of biology is playing down some research discoveries because they are politically incorrect or inconvenient? What if facts are being muted simply because people would prefer not to hear them? And – the biggest discomfort of all – what if some of our current medical advice is unjustified?

Impossible? Not really. It's happened before. True, the blaming of disease on evil spirits or bad humours to be released by sacrifice, bloodletting or leeches is far enough in the medical past to seem simply funny. Mention thalidomide, though – conjuring up images of babies with vestigial arms or legs thanks to a drug prescribed for morning sickness – and humour disappears. If such a major mistake can have been made so recently, what are the chances that mistakes are also being made now? Hopefully, low – but, for example, how good is the evidence that sunscreens protect us from skin cancer?

1

What If Sunscreens Cause
Skin Cancer?

It's a nightmare thought – and surely impossible if we judge from the enthusiasm with which those who should know urge all white-skinned people to use sunscreens when venturing out into the sun. In Queensland, Australia – current skin cancer capital of the world – intensive public campaigns urging the use of sunscreens have been active since 1983. On Bondi Beach, young entrepreneurs with spray guns squirt protective lotion on sunbathers who have arrived without their own. Even on some lukewarm British beaches, skin-cancer patrols – the first known as Operation Molewatch – dole out sun creams and advice to unprepared bathers. Weather-forecasters also do their bit and warn everyone of the risk from the sun's ultra violet (UV) rays in the day to come. Some children's clothing is marketed with spots which change colour according to sun intensity and sun Buster suits give neck to knee cover. Even the hoardings that advertise sun lotions now stress their protective as well as their tanning properties.

Such zealous actions have arisen from major sun-awareness campaigns launched by mainstream medicine. Doctors' surgeries all over the white world have leaflets and posters urging patients to be sensible when exposing their skin to the sun. Health authorities urge schools to reschedule outdoor activities away from the middle of the day and to provide sunscreen lotions and shaded playgrounds to help prevent pupils developing skin cancer. And the Imperial Cancer Foundation offers the following specific advice: wear a wide-brimmed hat that shades the face; if you work outdoors, try to reduce the amount of time spent in the sun at other times; avoid the sun between 11 a.m. and 3 p.m.; and wear a sunscreen of at least SPF 15 and a four-star UVA rating on skin that cannot be covered.

The hype and advice is everywhere – and Caucasians around the world are taking notice. But how good is the science behind the advice? What really drives the sunscreen steamroller – evidence or industry?

Rise and Rise

The logic behind sunscreen use seems reassuringly impeccable. Skin cancer is thought to begin when UV rays from the sun penetrate the skin and damage the DNA of the skin's cells, producing mutant genes that cause vulnerable cells to proliferate. The rationale of prevention is simple: stop the deadly rays from reaching the sensitive cells in the first place. Either don't go into the sun, stay clothed, or use a chemical sun block that screens out UV rays, allowing only harmless sunlight to penetrate to the vulnerable cells.

In the face of such powerfully simple logic, how could we question that sunscreens are anything but helpful? Yet, if we pause for a moment, there is a niggling worry. Sunscreens are, after all, manufactured chemicals that wouldn't normally cover the human body while it was exposed to the relentless glare of the sun. And time and again research has identified the trigger of cancer to be an unnaturally intensive and/or extensive exposure to some chemical or another.

As early as the eighteenth century it was noted that chimney sweeps were at an increased risk of scrotal cancer due – it was eventually discovered – to the way human skin transformed the otherwise harmless hydrocarbons in soot into carcinogens (cancer-causing agents). Since then, the list has grown and grown. Cigarette smoke – or the inhalation of large amounts of asbestos or coal dust, etc. etc. – can trigger lung cancer. Chemicals injected into meat-producing livestock may trigger bowel cancer in consumers. Excessive alcohol can trigger liver cancer. Cigarette smoke and excessive alcohol can trigger throat cancer. Excessive nicotine juice – from chewing tobacco leaves – can trigger tongue cancer. And so on – chemical, chemical, cancer, cancer. So where can we get reassurance that sun blocks actually prevent skin cancer rather than cause it? The obvious place to look is at how effective the use of sunscreens has been at reducing skin-cancer rates.

The first major campaigns urging white-skinned people to use sunscreens began in Australia and then the United States and Europe in the early to mid-1980s. And the campaigns were certainly powerful. By the early 1990s, sales of sunscreens in Australia were rising by 30 per cent each year. In 1993, $50 million Australian (£23 million) was spent on sunscreens; over half a million litres were rubbed onto white Australian skins. Similar increases occurred all over the fair-skinned world.

Yet, despite the obvious success of the sunscreen campaigns in winning public compliance, there is still no sign of the desired medical effect. All over the world, the incidence of skin cancer among fair-skinned people has rocketed throughout the 1980s and 1990s, roughly doubling every ten years. In Britain in 1935, the chance of a person getting the most dangerous form of skin cancer was one in 1,500. By the year 2000, the chance was one in 75 – a twenty-fold increase in risk – with no sign of the escalation slowing down, let alone reversing. As more people use sunscreens (and the more fervently they use them), the more are treated for skin cancer. What more evidence of cause and effect do we need?

The answer – of course – is a lot more. Just because there is a strong correlation between sunscreen use and skin cancer does not necessarily mean that sunscreens actually cause skin cancer. Maybe there are other ways of explaining the trends of the past twenty years or so – or maybe there aren't. There is a lot to consider before we can decide, but even then the conclusion isn't pretty.

Two Steps to Skin Cancer – and UV Blamed for Both

Skin cancer – like all cancers – develops as a result of abnormal cell reproduction. There are different types of skin cancer, depending largely on which cells in the skin begin to behave abnormally, and three of them – to be described later – concern us here. These are: Basal Cell Carcinoma (BCC), Squamous Cell Carcinoma (SCC) and malignant melanoma. They behave very differently from each other and pose different threats to a victim's survival. The main similarities are that they are all cancers of the skin, are all currently thought to be triggered by overexposure to the UV components of sunlight and are all thought to develop in a two-step process.

The first step occurs when something causes DNA in the centre of a cell – the nucleus – to mutate, changing the cell's character from obedient servant to malicious anarchist. Healthy cells divide only when told to by chemical signals reaching their DNA. And normally they reproduce at exactly the rate required to replace the naturally dying cells around them. Mutant cells have the potential to disobey the growth-regulating signals and to multiply anarchically.

In fact, most such mutations in the skin are probably short-lived because a damaged cell responds by producing an enzyme that chemically repairs such damage. We know that this must be happening because people with a defect in the gene that organizes such repair – a condition known as Xeroderma Pigmentosum – are 1,000 times more likely to develop skin cancer than other people. Even if the damaged DNA isn't repaired, though, cancerous cell multiplication still doesn't happen immediately; the first mutational step alone isn't enough to lead to skin cancer. For a long time after the initial damage, the body's defences – the immune system – seem able to keep the potentially anarchistic cells under control.

It is the second step that marks the real beginning of skin

cancer – immune suppression leading to a weakening of the body's control over the cell. Then the disobedient cell begins to divide and gradually forms a mass called a tumour or neoplasm. Even then, some tumours are benign and simply stay where they are, not spreading to surrounding healthy tissue. Others, though, are malignant – invading, compressing, and eventually destroying surrounding healthy tissue. Some of the cells can break away from a malignant tumour in a dangerous process called metastasis. These cells are carried by the blood or lymph systems to other parts of the body, where they continue to multiply and so form secondary tumours. Escape from immune control could be the reason why a high incidence of skin cancers is one of the commonest symptoms of immune deficiency syndromes – such as AIDS. It's a sobering thought that we could all be walking around with mutated cells in our skin, just waiting to become cancerous as soon as there is a dive in our immune systems.

According to current theory, both of the steps on the road to skin cancer are due to overexposure to UV light. Sunbathing and similar activities are thus thought to be doubly dangerous; not only might the UV light we expose ourselves to cause the initial mutation of DNA, it might also suppress the immune response. Exactly how UV light might cause DNA to mutate is unclear. It could be direct – the UV itself damaging the DNA. Or it could be indirect, the UV changing the skin's chemistry so that one of the products then alters the chemistry of the DNA. It is also unclear how UV light could suppress the immune system.

The evidence that UV light might suppress the immune system comes from experiments on mice in which animals exposed to UV light were less able than controls to reject tumours grafted on to them from other mice. Part of the problem in identifying the suppression process any further is that the mechanism by which the immune system keeps potential tumours in check in the first place is unclear. It may work like this: special messenger cells in the skin pick up fragments of proteins from the mutated cells, just as they do from other infectious agents such as viruses and bacteria. They then carry those fragments to the lymph nodes, which respond by dispatching killer T-cells specific to the problem. Enough killer cells are thus produced to keep the attacker or tumour in check. So perhaps UV light reduces the number of messenger cells, causing fewer defensive T-cells to be produced. UV light might destroy the messenger cells directly, or again it might simply trigger chemical changes in the skin that in turn kill or interfere with the messenger cells. One suggestion is that UV might alter the structure of one of the commonest chemicals in the skin, Urocanic acid, which in experiments on mice appears to be immunosuppressive.

Whatever is going on during the tug of war between tumour and immune system, there is some evidence that even growing tumours can sometimes be eradicated naturally. A tiny fraction of human skin cancer cases heal spontaneously. Nobody quite knows what happens on these occasions, but at least in mice, diet might be a factor. In tests, a low-fat diet seemed to curb the spread of cancer – and so, too, did drinking green tea! Mice exposed to UV and given green tea to drink instead of water reduced their risk of skin cancer by up to 87 per cent. The prime example of an ability to reverse skin cancer, though, is the Vietnamese pot-bellied pig. Work at the Royal Veterinary College in Britain found a particularly high incidence of skin cancer, shown in black patches, among its stock. However, after a number of weeks the cancer spontaneously disappeared in 99 per cent of cases, even if a pig initially had hundreds of tumours over its body. The pigs' immune systems had produced enough antibodies to destroy the tumours, leaving nothing behind but a white mark.

UV and the Different Types of Skin Cancer – Melanomas and Non-Melanomas

Human skin is like a two-layered cake sitting on a cake board. There is icing sugar on the cake's surface and jam in its middle.

The icing sugar on this skin-cake is a layer of dead cells – keratinocytes – that are continuously being sloughed off. The top layer of cake (the epidermis) is about as thick as a sheet of paper. It also consists mainly of keratinocytes but these are alive, not dead; a mixture of young, middle-aged and dying cells on a month-long journey from their birthplace in the jam (the basal layer of the epidermis) to their programmed death at the skin's surface. Although the cake board (the subcutaneous tissue) and the bottom layer of cake (the dermis) are many times thicker than the epidermis, they do not really play a major role in our story. We are much more concerned with the basal layer, which is where cancer forms, and the upper part of the epidermis, which tries to protect the basal layer.

The basal layer consists of a mosaic of pigment-producing melanocytes and column-shaped basal cells, interspersed with squamous cells whose primary function is to space and support the other two. These three types of cell are the most important progenitors of skin cancer. Basal cells continuously divide to produce an ever-changing layer of keratinocytes above them. In contrast, melanocytes do not continuously divide. Instead, they manufacture granules of the pigment melanin, usually in two different colours: brown eumelanin and reddish phaeomelanin. Most people produce both types of melanin but some redheaded people produce only phaeomelanin – a fact that seems to make them particularly vulnerable to skin cancer, as we shall see.

Types of Skin Cancer

The most common type of skin cancer forms from abnormal division of a basal cell and is known as Basal Cell Carcinoma (BCC). It begins as a small, pink lump that slowly enlarges. The next most common type forms from abnormal division of a squamous cell and is known as Squamous Cell Carcinoma (SCC). It begins as a thickening or lump that later breaks down to form an ulcer with a crust. Together, these two types of skin cancer are known as non-melanoma skin cancers, to distinguish them from the much more dangerous malignant melanomas which form from abnormal division of melanocytes.

BCC invades and destroys surrounding healthy tissue but it does not spread to distant parts of the body and so can easily be treated and is rarely dangerous. If left untreated, though, it can lead to quite severe disfigurement. SCC if left untreated can some-times metastase and lead to death. Together, the two forms are becoming increasingly common, with Australia leading the field. By 1990, 150,000 Australians needed treatment annually and the death rate was 1,000 per year. In Queensland alone, 30,000 or more new cases are being reported yearly; in the tourist region near Cairns around 10 per cent of men between the ages of sixty and sixty-nine have had carcinomas confirmed. In Britain there are roughly 45,000 new cases every year and about 500 deaths, with the number increasing yearly.

The much more dangerous variant of skin cancer, malignant melanoma, may start as either a flat or a raised mole. Many people have numerous moles on their bodies and deciding which might need closer examination and which are completely harm-less can be difficult. According to the American Cancer Society, the key signs are asymmetry (most melanomas have an irregular shape and if a line were drawn in the middle, one half would not mirror the other), a border (most do not have a smooth edge but have an irregular border which may also be inflamed and red), the colour (most are not just one colour but a mixture, including brown, black and blue) and the diameter (most normal moles are smaller than the blunt end of a pencil). In particular, it is the combination and rapid change in these signs that might indicate a melanoma. Other symptoms include itching, bleeding and satellite lesions – small marks such as a red area, brown spots, or a white ring – around the edge of the mole.

Malignant melanomas are dangerous because they can grow and metastase quickly. Surgery is effective if the cancer is detected early but once metastasis occurs, most treatments fail. Only 6 per cent of people live five or more years after diagnosis of advanced melanoma. By 1992, 7,000 Australians per year were acquiring melanomas and the rate was roughly doubling in each decade. In the United States in the 1980s the incidence among the white-skinned population increased by about 4 per cent per year. Then, through the 1990s as in other countries, the rate of increase was about 7 per cent per year. In Britain, cases doubled between 1980 and 1995 and by the year 2000 there were about 5,000 new cases each year from which about 1,500 people were dying; about a tenth as many cases as non-melanomas but with three times the deaths.

A few cases of melanoma – though perhaps no more than 10 per cent – are due to a genetic predisposition (with the probable location of the defective gene being on the short arm of Chromosome 1). The chances of a person who possesses this gene actually developing a melanoma are high – around 90 per cent.

Sun, Sex and Age: Body Parts Exposed

Current advice on how to prevent all three of these forms of skin cancer assumes that UV light is the culprit and that sunscreens are beneficial, not dangerous. This assumption comes largely from surveys showing who and which parts of the body are most at risk to the different cancers. The pattern of risks, though, is different for the different types. For example, melanomas strike all age groups but non-melanomas strike older age groups more than younger. Everywhere, non-melanomas erupt mainly in the over-fifties – though a recent trend in Australia has been for increasing numbers being reported by people in their thirties. Secondly, melanomas strike men and women equally whereas non-melanomas are twice as likely to strike men than women. In 1995, for example, 697 men and 698 women died of melanomas in England and Wales. By comparison, fourteen Queensland men in every 1,000 acquire non-melanomas each year compared with only seven women (in the southern states of the United States the comparable figures are six and three). This apparently greater vulnerability of men to non-melanomas, though, may actually be an artefact – because a third factor is that non-melanomas are most common in people who work outside. Melanomas, on the other hand, are more common in people who work indoors, particularly office workers. It may simply be that more men work outside whereas the numbers of men and women are more evenly split in offices.

This link between working outside and a greater risk to non-melanomas is one of the reasons that sunlight is thought to trigger BCCs and SCCs. Another is that the tumours on both sexes occur most often on the most exposed parts of the body – the face, hands, arms and neck – and become increasingly common the nearer to the equator that white-skinned people live. The lighter a person's skin and the more freckles they have, the greater they are at risk to non-melanomas. Redheads are ten times more likely than the average person to develop non-melanomas. Other people whose skin is so light that they burn rather than tan are 50 per cent more likely to develop non-melanomas than people who tan rather than burn. In one study, people reporting just one case of sunburn during childhood with pain lasting two or more days were more likely to develop the dry, scaly growths of skin called solar keratoses which are a precursor for SCC. And people reporting six or more painful sunburns in their lifetime were three times more likely actually to develop SCC. The current theory of the generation of non-melanomas, then, is that sunburn during childhood triggers mutations in the exposed parts of people's skin but that the cancer does not then develop for another twenty – perhaps even forty or fifty – years. The critical factor affecting development is thought to be the accumulation of damage due to total exposure to sunlight. The more and the stronger the sun to which any given patch of a person's skin is exposed, the greater the risk that a non-melanoma will develop on that patch.

A quite different theory, though, is needed to explain the pattern of attack by melanomas. Not only do they strike most at people who work indoors, they also strike at parts of the body that are less exposed than the face and hands: the trunks of men and the legs of women. Melanomas are also unusual among cancers in that they tend to strike affluent people more than less affluent people. The favoured theory – and the one on which current advice to prevent melanomas is based – is that melanocytes are vulnerable to intermittent exposure to sunlight strong enough to lead to burning. The most vulnerable time of life is thought to be during childhood. This is because people who migrated to Australia when younger than ten years old have been found to suffer the same incidence of melanoma later in life as people who were born in Australia. But those migrating to Australia after the age of fifteen years subsequently suffer only the same incidence as people in their homeland. The implication – as with non-melanomas – is that it is sunburn during the first ten years or so of childhood that determines the risk of melanoma twenty or so years later in life. It is even thought that only one to three episodes of blistering sunburn in early childhood might be enough. Intermittent exposure alone is thought to increase the risk of malignant melanoma by 70 per cent, while sunburn might nearly double the risk. Perhaps children in the most affluent families are most affected because they are the most likely to fly off to the sun for a beach holiday for just a couple of weeks each year, a habit that they continue into an adult life spent largely working indoors.

The Logic of Natural Protection

Most mammals are covered with hair, move in and out of the shade, and have protective melanin deposits in exposed areas of skin, such as the tips of their ears or their noses. Ravages from UV light and skin cancer are not a significant problem. When skin cancer is reported in mammals outside the laboratory, it usually concerns domesticated species – bred artificially for characteristics other than resistance to UV light – rather than wild animals. In particular, skin cancer has been noticed in species transported from temperate latitudes to Australia. Those at greatest risk are all without pigment – white cats, white-faced Hereford cattle and English bulldogs – and it is mainly their ears and noses that are vulnerable to the searing Australian sun.

When our human ancestors first evolved, losing their protective layer of hair, natural selection changed their skin so that melanin production occurred more or less all over the body, thus continuing to provide protection. The main mechanism is as follows. Although melanocytes produce melanin granules within their own cell body, these granules of pigment are soon ejected and shared among the surrounding newborn keratinocytes. As the keratinocytes migrate outwards towards the skin's surface they carry the pigment with them – as if raising an umbrella over the cells they leave behind.

Perhaps surprisingly, all people have about the same number of melanocytes – usually between 1,000 and 2,000 per square centimetre. Variation in skin colour between races and individuals depends much less on differences in the number of melanocytes than on differences in the types and amount of pigment they produce. The darkness of a person's skin depends on the size and number of the melanin granules that each melanocyte produces; the hue of their skin depends on the relative proportions of eumelanin and phaeomelanin. Tanning occurs when the melanocytes in the skin of a normally white person increase their production of melanin.

Melanin and the process of tanning are prime examples of the exquisite nature of natural selection. Melanin does more than screen out UV light; it is selective over what it does or does not screen and the melanocytes respond to what gets through by adjusting their level of melanin production. This is necessary because the sunlight that strikes the outer layers of the earth's atmosphere has three main bands of UV – usually known as UV-A, UV-B, and UV-C – in its spectrum. The different bands have different characteristics and pose different threats. Physically, UV-C has the most energy and potentially could do the most damage to exposed skin. However, a layer of ozone in the earth's stratosphere – to be discussed later in connection with global warming – is relatively impenetrable to UV-C and prevents most of it from reaching the earth's surface. UV-B is intrinsically less damaging than UV-C but, although much is blocked by the stratospheric ozone layer, enough still gets through to be dangerous. UV-A is an even lower energy band of light but is blocked even less by the ozone layer. Purely in terms of energy, it should take about 1,000 times as much UV-A to damage skin as UV-B. But as ten to a hundred times more UV-A passes through the atmosphere than UV-B, the difference in threat from the two sources is not as great as it might seem – only a factor of ten to a hundred.

Natural selection adopted melanin as the sun-screening pigment for human skin because it absorbs UV-B and UV-C but allows through UV-A. This is not an evolutionary oversight but a clever part of the tanning system by which the skin protects itself. In effect, the skin monitors the amount of relatively harmless UV-A penetrating through to the basal layer. It then adjusts the production of melanin just enough to screen out the UV-B and any UV-C. The more UV-A that penetrates the skin, the more melanin is produced to screen out the higher doses of UV-B that must also be arriving.

A suitably dense layer of melanin in the epidermis is still undoubtedly the best natural protection against UV light for humans. Skin cancer is a white-skin affliction. It is rare in dark-skinned people – even those living naked in the tropical sun – and uncommon in white-skinned people who tan easily. People who tan rather than burn suffer 33 per cent fewer non-melanomas than people who burn rather than tan. And redheads lacking eumelanin completely are ten times more likely than average to develop a non-melanoma. This could be due to the differing amounts of UV light that penetrate people's skin – but there could be another explanation, as we shall discuss later.

So why did natural selection make the apparent mistake of producing people who lacked the permanently dense layer of melanin that would have given them sure-fire protection? Why go to the lengths of evolving such a complex procedure as tanning when a black skin would be safer? There can be only one logical answer – a little UV-B penetration is beneficial, too much is bad. Screening out too much UV-B must actually be detrimental to our health, just like screening out too little.

The main benefit of sunlight on the body has been known for a long time. It is necessary for the stimulation of the production of vitamin D, which is essential for the development of healthy bones. In addition, it improves skin conditions such as psoriasis and makes people feel psychologically better, perhaps reducing the frequency of depression (particularly the variant known as SAD, Seasonal Affective Disorder). It is even possible that sunlight reduces the risk of coronary heart disease and multiple sclerosis, again perhaps through its effect on levels of vitamin D.

Too much melanin in the skin can hinder the production of vitamin D. People from the tropics, with a naturally dense layer of melanin in their skin, can suffer badly from vitamin D deficiency when they migrate to live in higher latitudes such as northern Europe and North America. The weak temperate sun just does not have the strength to penetrate the melanin and stimulate vitamin D production. Skin diseases are common and so, too, are bone-development illnesses, such as rickets. Vitamin D supplements to the diet are often necessary if such migrants are to remain healthy.

Without vitamin supplements to hand, evolution did the next best thing when our ancestors began to migrate from their African cradle to colonize less sunny climes: it reduced the level of eumelanin production as appropriate to the strength of the local sun. And in the case of people entering cloudy climes, it often reduced production completely, producing, for example, the redheads of the Atlantic coast of north-west Europe. Many of the areas colonized, though, were seasonally sunny. What was needed was a tanning system that increased the production of protective melanin as the sun grew in strength, followed by decreased production when the sun grew weak.

Unfortunately, of course, in many climates, the sun does not pass slowly and smoothly from strong to weak as the seasons progress. Often, long periods of cloud can suddenly give way to days of clear skies and intense radiation. And increasingly as our species migrated away from the equator, people wore clothes that could be put on and shed according to the weather. This produced yet further problems as evolution tried to balance vitamin D production and UV protection. In our evolutionary past it was inevitable that on occasion skins without a protective melanin layer would be bombarded by UV-B and would burn and be damaged. Evolution needed one final solution – and the one it produced was peeling, the mechanism by which it sheds any layer of damaged cells to replace them with new and undamaged cells. If the damaged skin does not burn and peel, the damaged skin cells can survive and reproduce, potentially leading to cancer. There may even be an additional protective mechanism – moles. Quite possibly these are localized areas of heavy pigmentation to protect 'hot spots' on a person's skin; positions at which sunlight has been particularly focused in a person's past and which may need special protection in the future.

So melanin, tanning, peeling and maybe even moles are evolution's solutions to the twin dangers of too much UV light and too little vitamin D production. It leaves us with two big questions: how complete were these solutions at the time, and how robust are they for modern humans? We can't really answer the first question – but the optimist's view would be that the solutions were perfect. Consider Negroes in the tropics, redheads in northwest Europe, blondes in northern Europe and dark-haired people in southern Europe, etc. Suppose that they were all naked from birth and were free to move in and out of the sun as they felt the need. Suppose that they travelled only by walking. And suppose that they never applied chemicals to their skin. How many would suffer vitamin D deficiency or skin cancer? We don't know, of course – but it would be a fair guess that only a few would do so.

But this isn't how we live our modern lives. Global migration has taken many a sun-adapted Negro to the gloom of the temperate zone and many a shade-adapted redhead to the tropics.

People are rarely naked from birth, so the tanning process is unable to weave its protective magic in the exquisite way it was intended. Nor do people's occupations allow them to move in and out of the sun as they feel the need. Some are forced to work day in, day out in the artificial light of offices; others are forced to work long hours in the relentless sun. Air travel can wrest a person from sombre winter skies to tropical sunshine in a matter of hours. And people's skins are continuously exposed to a mixture of pollutants, soaps, cosmetics, lotions – and, of course, sunscreens.

It isn't at all surprising – with such a difference between our evolved state and modern life – that skin cancer is on the increase. The question, though, is whether the whole package of twenty-first-century life is to blame or whether in this extensive list there is one main culprit.

The Logic of Artificial Protection – Sunscreens and Solaria

Sunburn is unpleasant. It happens when the density of melanin in our skin isn't adequate to screen the underlying tissue from the cooking effect of UV light. From time to time even our naked or near-naked, free-living ancestors will have sunburned. Although they, better than us, will have been able gradually to build up and lose their tans as season followed season, they will still have been vulnerable to sudden changes in the weather.

We stand even less chance of being prepared for the first hot and sunny days of spring and summer. Yet, when they arrive, white-skinned people are drawn to the outdoors like a magnet, throwing off their clothes and soaking up the sun's rays. Whether as slaves to fashion or slaves to some basic urge, such people enjoy the sun and consider a good tan to be a sign of health, vigour, attractiveness and affluence. But on a day-to-day basis, few of us can strip when the sun shines during the lukewarm days of spring, thereby building up a slow and effective tan in advance of the really hot days of summer. Many of us are forced to work indoors for fifty weeks a year, then in desperation fly in search of hot sun for the remaining two. Or we are forced to work outdoors in burning sun far beyond the stage that, if free to do so, we would seek shade. Whatever our personal circumstances, an annual sunburn is a real possibility. There is a demand, therefore, for some artificial aid. Something that will allow us to indulge our urge to be in the sun by not only protecting us from sunburn but also allowing our skin to increase its melanin production to the level needed for natural protection. The more we globetrot, the greater the demand. And science has come up with two such aids – the chemical sunscreen and the solarium.

The ideal sunscreen was initially thought to be one that exactly mimics the action of melanin. In other words, it screens out the dangerous rays of UV-B and UV-C but allows through UV-A. That way, it protects us from sunburn while at the same time allowing melanin production to be stimulated so that we eventually tan. Sunscreens are usually marketed with their ability to screen out UV-B described in terms of a sun protection factor (SPF). SPF is a measure of how long it will take the sun to redden skin protected by the screen compared with how long it would take to redden unprotected skin. An SPF of 8 means it takes eight times as long for redness to appear, which should mean that the screen is blocking out nearly 90 per cent of UV-B light. For comparison, lightweight clothing has an SPF of between 5 and 15. However, countries differ in the way manufacturers test the SPF factor of their products. For example, Australia carries out tests using volunteer humans exposed to lamps designed to simulate the UV spectrum of the sun. In Germany, the lamps used don't attempt to simulate the solar spectrum, and in Britain some manufacturers don't use humans to test their products but use mice instead.

In recent years, with concern growing that UV-A might not be as harmless as was first thought, sunscreens have changed and increasingly block UV-A as well as UV-B and UV-C. A separate rating system has been introduced for UV-A – the star system. These stars, though, are not a straight indicator of how much UV-A is blocked, but rather a measure of how much UV-A is blocked relative to UV-B. Thus a four-star rating on a factor 15 sunscreen is different from a four-star rating on a factor 5. There have been recent moves to try to make the UV-A rating more explicit.

Another approach used by increasing numbers of people trying to avoid sunburn on holiday is to build up a tan in a solarium before exposing themselves to the rays of the real thing. This has the additional advantage, of course, of giving the person the cosmetic benefit of a tan at times of year when tans are uncommon. Tanning salons in the United States doubled in number every two years in the late 1980s and now over 1 million Americans use commercial solaria each day. In Britain, the figure is more like 3 million people per year.

Proponents argue that obtaining a suntan in a controlled environment through the sensible use of a sunbed is a more responsible way to get a suntan than overexposure to natural sunlight. Opponents, though, argue that the practice is dangerous. The main problem with artificial tanning has always been the difficulty of producing lamps that accurately mimic solar strength and spectrum. Early versions of tanning lamps produced large doses of UV-B and often burned the sunbathers. Some commonly used mercury halide lamps have five times the UV-A found on a sunny day at the equator. Any deviation from natural sunlight takes the artificial light away from the range that the human tanning system has evolved to respond to safely and effectively.

In experiments on genetically hairless mice, some of the early tanning lamps produced 100 per cent skin cancer; even some of the common modern lamps produce cancer in 5 per cent of mice. There are reports from the United States and elsewhere of people dying from using sunbeds. The number is small – probably only twenty worldwide – but if skin cancer takes decades to develop, it could be a while before the full impact of solaria is seen; the sunbed industry is still relatively young. Forecasts have been made that between 50 and 60 per cent of people who use sunbeds once or twice a week for four years would develop precancerous cells that could eventually lead to cancer.

Do Sunscreens Work?

There is absolutely no doubt that sunscreens help to prevent sunburn and peeling. After all, their effectiveness is measured by how much longer it takes skin to burn with sunscreen as compared to without sunscreen. But do sunscreens prevent skin cancer? There is no clear answer.

On the positive side, some experiments exposing mice to sunlamps have reported that sunscreens seem to prevent at least some mice from getting cancer. Studies of humans, too, have provided some indirect evidence that they might be protective in that regular sunscreen use can prevent solar keratoses – discoloured spots that are a precursor of SCC and a risk factor for melanoma. In a study, fifty people who already had solar keratoses used broad spectrum UV-A and UV-B (SPF 17) screen every day for a summer. By the end of the summer, they had fewer new keratoses than a control group using a non-sunscreen cream. They also had more remissions, old keratoses disappearing. On the negative side, experiments in which sun lotions were rubbed on the ears of mice before exposing them to a sunlamp showed an enhanced growth of melanomas. Another study found that people who used sunscreens in fact suffered more melanomas.

So, with the evidence that sunscreens might protect against cancer being at best confusing, we can return again to our original question. Might sunscreens actually cause the cancers they have been designed to prevent?

What Causes Skin Cancer – UV Light or Sunscreens and Other Chemicals?

UV light undoubtedly has the power to cause cancer in the skin of mammals. As early as the 1930s and 1940s experiments on mice showed that doubling the dose of UV halves the time to the onset of a tumour. UV light – particularly UV-B and UV-C wavelengths – is dangerous. This is why our ancestors evolved whole-body melanin and a tanning system when they lost their hair and strode naked through the African sun en route to colonizing the world. By the same token, though, despite the ever-stronger finger of suspicion, we should not intuitively expect UV-A to be dangerous to human skin. If it were, natural selection would not have devised a screening system deliberately transparent to these wavelengths.

We have to ask, though, whether experiments on the skin of mice really tell us anything about humans. Before mice were brought into the laboratory and turned into myriad genetic strains, including the hairless mouse, the species was active by night, skulked in the shade by day, and was covered with hair. Evolution had never needed to protect them from high doses of UV light in the way it had humans. When an experimenter turns his UV spotlight on a shaven or hairless mouse, relative to human skin he or she is dealing with skin that is unprotected and unprepared in all senses of the word. Not only are laboratory studies usually carried out on artificially defenceless mice, they invariably use sunlamps rather than natural sunlight, thereby distancing the experiments still further from relevance to humans in the sun. And to emphasize the point, research into melanomas has been particularly hampered because no one has yet been able to make natural sunlight cause a melanoma in a laboratory animal.

Suppose, for the moment, that UV light were not the cause of the skin-cancer epidemic we are seeing in humans. Suppose we step back from the hype, propaganda and advice of the past two decades and ask what else might be to blame. As in any consideration of cancer, the first and obvious place to look is our environment of man-made chemicals. Particularly as we now know our skins are virtual sponges to whole ranges of chemicals. For many years, it was thought that human skin was an impervious layer that protected us fairly completely from our chemical environment. We now know that this isn't the case. Studies using human tissue, such as removed during mastectomies and other surgery, have shown that most chemicals we put on our skin are absorbed to some degree. They have also shown that human skin has very different absorbent properties from rodent skin, casting doubt still further on any conclusions reached from research using laboratory animals. Such absorbency means that any one of the range of soaps, talcs, aftershaves, lotions, medications – and of course sunscreens – that we have been putting on our skin over the past few decades could have been absorbed into our tissues, thereby triggering illnesses. For example, in the 1970s hexachlorophene was used as an antiseptic in baby soaps and talcs but was later blamed for causing brain damage and death in some babies after penetrating their skin.

As in the case of hexachlorophene, most of the effects of chemicals absorbed through the skin would have been apparent relatively quickly. But the effects of chemicals that cause skin cancer are likely to be delayed, to surface years or even decades later. Even so, some industrial chemicals have already been associated with an increased risk of skin cancer. These include coal tar, soot, pitch, asphalt, creosote, paraffin wax, petroleum derivatives and arsenic. As a result, the more obvious occupational skin cancers are declining because most people at risk now wear protective clothing. Nevertheless, in Britain professional chemists are still at an increased risk to skin cancers – and in Taiwan and perhaps also in Mexico and Germany, risk of skin cancer correlates with the level of arsenic in the water supply. A chemical threat to our skin is clear.

So, if it is correct that events during childhood are particularly to blame for melanomas and non-melanomas twenty or more years later, how can we be certain that the current skin-cancer epidemic among people fifty years old or more is not due to chemicals that were rubbed on their skin when they were children in the 1950s and 1960s? It is not difficult to find examples of skin-care products that in hindsight were risky. For example, we have already seen that Urocanic acid has been linked to the immune-suppression step in skin cancer. Long before this was suspected, the chemical was used as a moisturizing agent in a whole range of skin lotions. In the 1960s, it was even hailed as a natural sunscreen – and was still present in some skin products as late as 1991.

As a further example, a study of a well-known range of tanning lotions in the mid-1990s was carried out on nearly 1,000 people. One group that had developed melanomas was compared with a similar but otherwise healthy control group. The study found that fair-skinned people who had used sun creams containing the chemical psoralen, which is extracted from bergamot oil, were four times more likely to develop malignant melanomas than people who had not.

The people taking part in the study were recruited from France, Belgium and Germany because these were the main countries in which the sale of sun lotions with such high concentrations of psoralen was allowed. Earlier studies on rodents had already suggested that the main ingredient might be carcinogenic, so the products had never been approved in the United States. They were also banned in Switzerland – and in Britain could be sold only if the concentration of psoralen was less than one part per million (thirty to sixty times lower than allowed in the countries studied). After the study was published, a European Union committee voted to limit the concentration of psoralen in cosmetics to one part per million. Predictably, the manufacturers complained. So, too, did French health ministers and various scientists. Their main objection was that any study that relies on people remembering what products they have used over the years has its limitations. And of course, it's always possible that the suspect lotions were simply not as good a sunscreen as the lotions used by the healthy group.

Maybe the lotions containing psoralen were not triggering cancer; maybe they were simply not preventing it. But psoralen is not the only suspect chemical in sunscreens. One of the commonest UV-B blockers is Padimate-O. In sunlight, this chemical may generate free radicals known to attack the body's DNA and so in principle could easily cause the mutations that lead to skin cancer. So far, then, there are just as many grounds to blame skin cancer on chemicals as there are on exposure to the sun. The final test, though, is whether this alternative theory explains the pattern of risk to skin cancer as well as the UV theory on which all our advice is based.

Consider non-melanomas first. If we are searching for a chemical cause to explain the observed pattern of risk, we are looking for something – or a range of things – that comes more into contact with the skin of people who: are white-skinned; live nearer the equator; work outside; burn rather than tan; and experienced bad sunburn episodes during childhood. The chemical must also have been used more and more over the past few decades, particularly on parts of the body – the face, arms and hands – that are most exposed. There must be many possible candidates, but the most obvious is sunscreen, the cumulative use of which from childhood onwards eventually frees the non-melanoma tumours once a person reaches forty or fifty. We cannot rule out the possibility, though, that sunscreens rather than acting alone are interacting chemically with some other substance – perhaps something as mundane as soap or as esoteric as mosquito repellent – in producing a final, carcinogenic effect.

Now consider melanomas. If we are searching for a chemical cause to explain the observed pattern of risk, we are looking for something – or a range of things – that is used: intermittently on the skin of children, particularly the trunks of boys and the legs of girls; more on children who burn rather than tan; more in Australia than elsewhere; more by more affluent families; more on children destined to become office workers; and again has been used increasingly in recent decades. Once again, there must be many possible candidates, but the most obvious are from among the ranges of sun-care products.

Maybe the differences between melanoma occurrence and non-melanomas are due to different parts of the body responding differently to the same chemical – or to different parts of the body responding differently to different chemicals. Maybe non-melanomas are more likely to be triggered by, say, tanning lotions or sunscreens, and melanomas by, say, sun aftercare. Or maybe it is simply that different people respond differently.

If there were a chemical explanation for the recent epidemic of skin cancers, how would it work? Probably little differently from the way UV has been thought to work. The culprit chemical (or chemicals) could, after penetrating the skin, cause the mutation of DNA in the first place. Or it could interfere with the DNA repair system. Or it could suppress the immune system. Or it could interfere with the burning and peeling process that evolution designed to rid the skin of potentially cancerous sun-damaged cells. Or it could do all four. Alternatively, something else could trigger the initial mutation – an insect bite, for example, or even UV light – and the sustained use of the chemical offender could be the cause of the tumour running out of control instead of being held in check by our bodies' defences. All of these are possible from what we know at present.

We are faced therefore with a situation in which the available evidence simply isn't capable of telling us whether sunscreens are the sought-after protection or a potential menace. For as long as the use of sunscreens and the incidence of skin cancer continue to rise side by side, a worry remains that the former really is causing the latter and that the correlation isn't a simple artefact. Part of the problem is that none of the evidence for or against sunscreens is good evidence. Much of it is based on correlation. So, what if people living near the equator are at greatest risk? Is that because they experience more sun, use more sunscreen, or maybe even use more mosquito repellent? Redheads are also at greater risk. But is that because they are naturally less protected against UV light, or because they use more sunscreens, or because the particular chemistry of their skin reacts differently to any number of chemicals?

The only real experimental evidence comes from studies of rodents. But not only do such animals have a completely different set of skin defences and requirements from us, they develop their tumours within weeks rather than decades. Moreover, in most experiments they are exposed to artificial light, which may or may not mimic the subtle but important details of sunlight.

When the evidence does come from studies of people, they have usually been asked to remember things such as what sun-screens they have used for the past twenty years and how often they were sunburned as children. Health Authority statistics for Britain show that four out of ten children are sunburned each year. It is asking a lot for people past fifty to remember how often it happened to them. Would it be a surprise if those who had already succumbed to skin cancer remembered more instances of sunburn when children than those who had not, whatever the true number?

With the best will in the world, the sun-care industry cannot scientifically reassure its customers that this or that product is safe. If it is true that melanomas and non-melanomas take twenty or more years – maybe fifty – to develop, it is simply too soon for any product to claim that its cumulative use over many years is not associated with an increased risk of skin cancer. Even then, the only scientifically robust experiment would be one in which one group of people systematically used a particular product over their lifetime while another group unknowingly used a placebo from an identical bottle. Only if there were no difference in skin-cancer rates once the two groups were sixty years old could the manufacturers claim to have proved their product was safe. Clearly, such an experiment would be unethical as well as imprac-ticable. Yet, without such evidence, the advice being given can only be based on data that in most scientific enquiries would be considered flimsy, to say the least.

The Sunbather's Dilemma

So, what should we do if we wish to protect our children and ourselves from the risk of skin cancer? Should we avoid letting either sunlight or sun-care chemicals anywhere near their or our skins? Should we only expose ourselves to the sun if we protect ourselves with SPF 15 and four-star UV-A sunscreens? Or should we fearlessly indulge in sunbathing but avoid the temptation to use any chemical aid?

On the available evidence – when we step back from dogmatic advice and advertisers' hype over how to cut the risk of skin cancer – the unsettling answer is that there is no proven answer. The safest option – hedging all bets – is to use light, loose clothes and hats as initial protection rather than sunscreens. Ideally, a natural tan should be built up gradually and from an early age – using real sunlight, not solaria. Finally, chemical interference with the burning and peeling process – if burning inadvertently occurs – should be a last resort rather than a matter of routine. Above all, we shouldn't spurn or become afraid of the sun, because sunlight has known benefits as well as possible dangers.

The British government has set itself the target of halting, if not reversing, the current year-on-year increase in skin cancer by 2005. They may succeed if current theory and past advice is correct. If, however, chemicals cause skin cancer – even the very chemicals supposed to prevent it – the trend will not change until the advice changes.

2

Cholesterol: The Good, the Bad and the Ugly

Of all the dietary demons, cholesterol should surely be king; a greasy killer responsible for the heart attacks that kill one in four people in industrially developed countries. Throughout the twentieth century, cholesterol's handiwork – coronary heart disease – has ebbed and flowed across the western hemisphere like a plague. Huge amounts of money have been spent on unveiling the killer's work through science – and on the back of the results, national advisory councils have worked tirelessly towards educating consumers about cholesterol's evil ways. The National Cholesterol Education Program in the United States, for example, has set itself the goal of reducing the consumption of animal fat – the major dietary source of cholesterol – to about 10 per cent of total calories for all Americans.

People could be excused for thinking that such campaigns are based on solid science and have only the consumers' best interests at heart. But is cholesterol really such a villain? Will the lowering of cholesterol in a nation's blood really reduce the number of deaths from heart attacks? Just how robust is the science behind the attempted control of cholesterol in the human body?

The Nature of the Beast

Cholesterol is a fatty substance called a lipid. It is a steroid rather than a true fat; part of its molecule is an alcohol, though you wouldn't want it in your favourite cocktail. Some cholesterol enters the body directly in food, particularly in butter, eggs, brain, liver and fatty meats (poultry and fish are low in cholesterol and cereals, fresh fruit, and vegetables contain none). However, most cholesterol – around 80 per cent – is manufactured by the body from fat. Dietary fat comes in three main forms – polyunsaturated, monounsaturated and saturated – depending on molecular structure. Of the three, the most important for cholesterol production is the saturated form found in animal fats and some vegetable fats, such as palm oil.

Cholesterol's trademark is its insolubility in water, a useful trait that has led evolution to weave the chemical into cell walls whenever these need to be waterproof. Nerve cells, for example, employ cholesterol to isolate their insides from changes in their surroundings; the highest concentrations of cholesterol in the human body are found in the brain and elsewhere in the nervous system. But insolubility – in blood as well as water – raises problems of transport around the body. So, in humans, cholesterol is transported inside spherical particles composed of lipoproteins. Like a fleet of submarine carriers, these lipoprotein spherules carry cholesterol around the body, each one carrying about 1,500 molecules of the stuff on a journey that lasts an average of about 24 hours. Not all the carriers, though, are the same and their roles in cholesterol biology are very different. The most important are the heavy (HDL: High Density Lipoprotein) carriers and the light (LDL: Low Density Lipoprotein) carriers.

With most of the body's cholesterol being produced and processed by the liver, there is non-stop traffic of lipoprotein carriers between this busy organ and the other tissues of the body. Essentially, the light carriers ferry cholesterol away from the liver; heavy carriers bring it back to the liver. When any of the body's cells need cholesterol, they signal for a delivery by the light carriers, which dock and unload at special delivery bays – LDL receptors on the cell surface. Any cholesterol returned to the liver by the heavy carriers is either used – for example in the manufacture of certain hormones – or excreted in bile. At any one time, between 60 and 80 per cent of cholesterol in the blood is being transported by light carriers compared with 15 to 20 per cent by heavy carriers. The remainder – usually much less than 25 per cent – is being transported by other lipoproteins.

Most of the time, the cholesterol transport and processing system works smoothly as the carriers slavishly follow the routes and timetables devised for them by evolution. Cholesterol molecules are clearly ubiquitous and hard-working – but it is claimed they also have a dark side. When people suffer a heart attack – their coronary arteries strangled as oxygen cannot reach the heart muscles – massive gangs of cholesterol molecules are always at the scene of the crime. It is as if the transport system has run amok, the lipoprotein carriers anarchistically dumping huge amounts of cholesterol on an artery wall until all passage is blocked. No wonder scientists and the national committees they advise have pointed prosecuting fingers in the lipid's direction.

There is, though, a proviso – because not all cholesterol in the blood can be deemed bad. After all, the cholesterol being carried by the heavy carriers is actually being removed from the peripheral tissues and is on its way back to the liver for processing or disposal. This cholesterol, therefore, can be considered 'good' – but there is little of it. It is outweighed by a factor of three to five by the so-called 'bad' cholesterol being carried by the light carriers to the peripheral tissues. It is this 'bad' cholesterol that national heart foundations would dearly love to check in the human body.

Case for the Prosecution

The evidence that cholesterol is a killer is impressive. Not only is it always present in blocked arteries, but also people with a high level of cholesterol in their blood are more likely to suffer a heart attack than people with normal or low levels of cholesterol. This plain and irrefutable fact has emerged from scores of studies around the world – and when we look at each step in the sequence from diet to heart attack, the case against cholesterol seems watertight.

The Accomplice: A High-Fat Diet

The scientific investigation of cholesterol suggested very early on that the lipid had an accomplice in its attack on the human heart. The finger was pointed at dietary animal fat. Not simply the total intake of animal fat, though: the ratio of animal to vegetable fat – or, more specifically, the ratio of saturated fatty acids (SFAs) to polyunsaturated fatty acids (PUFAs). Finnish people, for example, eat extremely fat-rich food, much of which comes from animals, and have the highest blood cholesterol levels in the world at around 270 milligrams per decilitre (mg/dl). Vegetarians, on the other hand, tend to have lower cholesterol levels than other people – and by definition eat relatively little animal fat. In rural China in the 1980s the diet was low in fat and largely vegetarian, almost vegan; only 15 per cent of dietary energy came from fat and virtually all of that came from plants, not animals. Blood cholesterol levels were low – on average around 150 mg/dl. By comparison, in Britain in the 1980s between 40 and 45 per cent of dietary energy came from fat, with well over a third of that coming from animals; blood cholesterol levels averaged around 250 mg/dl.

When diet changes, so does blood cholesterol level. Laboratory tests on rodents have shown that an increased intake of PUFAs can lower blood cholesterol concentration – and a review of sixteen trials on humans around the world concluded that changing from a high animal-fat to a high vegetable-fat diet can produce a significant lowering of the blood cholesterol level. One of the most bizarre trials involved a group of eight people – four men and four women – who, in the early 1990s, sealed themselves inside a self-contained 'ecosystem' for two years. The sealed glass and steel structure was known as Biosphere 2 and was set on 1.2 hectares of desert just north of Tucson, Arizona. The biosphere was designed as a luxuriant greenhouse, complete with farm, desert, rainforest and mini-ocean and the volunteers' plan was to live solely off what they could produce in their own little world. And they stuck to it, despite overcast weather and losses due to insect pests greatly reducing the yield of the crops they were cultivating. As a result, they were forced to follow a strict low-calorie but highly nutritious diet. This consisted mainly of fruits, grains, beans and vegetables, supplemented with inevitably small amounts of goat's milk, goat meat, pork, chicken, fish and eggs. During the first year, the 'biospherians', as they came to be known, lost almost 10 kilograms each – and their cholesterol level fell from an average of 195 to 123 mg/dl.

It isn't only adults whose cholesterol levels drop while on low animal-fat diets. In a Finnish study, more than 1,000 babies aged seven months were divided at random into two groups. Half received a diet in which no more than 35 per cent of the energy came from fat, and in which saturated fat made up only one-third of the total fat. The other half received unrestricted fat, which in Finland means 40 per cent of total energy coming from fat of which half is saturated. At the age of thirteen months, children on the lower-fat diet had maintained their initial choles-terol levels yet had grown just as well as the other children. Those given a normal Finnish diet, however, had increased their blood cholesterol level by 8 per cent.

The Modus Operandi: Strangled Hearts

Cholesterol strangles hearts – and the more of it there is in a person's blood, the more effectively it seems to do so. Strangula-tion occurs by a process known technically as arteriosclerosis, or colloquially as 'hardening of the arteries' – because it involves the hardening, thickening, and loss of elasticity of the artery walls. Eventually, blood and oxygen cannot get through to the heart muscles and the organ effectively suffocates and stops.

Arteriosclerosis comes in several forms but by far the common-est is atherosclerosis, which affects medium and large arteries, especially those that carry blood to the heart, brain, kidneys, and legs. The first visible sign of atherosclerosis is when cholesterol begins to form deposits – fatty streaks – on the inner lining of the arteries. The sequence isn't yet fully understood but seems to begin with biochemical damage to the light lipoprotein carriers of 'bad' cholesterol. Macrophages – a type of white blood cell whose normal job is to mop up damaged or dead cells – devour these damaged particles and eventually become so engorged with cholesterol that they get stuck to the arterial wall. Over a period of years, these fatty streaks enlarge and thicken to form plaques with rough edges that irritate the smooth lining of the arteries. Cells die, smooth muscle cells multiply and accumulate, and scar tissue forms. Calcium also precipitates – about 96 per cent of people with heart disease have calcification of the coronary arteries. Platelets, the blood cells that initiate clots, settle and sometimes grab other platelets from the passing flow, eventually congealing on top of the plaque into a hard, pale clump called a thrombus.

When the process begins, fatty streaks are seldom deeper than 1 millimetre. Plaques, though, can be several millimetres deep and can cause dangerous narrowing of arterial bores. Eventually, the artery can block entirely. It can rupture, or a chunk of thrombus can break off and lodge in a smaller artery. Exactly what happens – and what symptoms occur – depends in part on a plaque's composition, for plaques vary in the proportion of lipid and fibrous material they contain.

The heart, like the brain, is particularly sensitive to having its blood – and hence oxygen – supply interrupted. If a part of the heart loses its blood supply, its ability to function as a pump is reduced within seconds and fails completely within minutes. This is the underlying cause of most heart attacks and is usually triggered by a plaque clogging up the artery serving a particular area of the heart. Plaques that are mainly fibrous seldom rupture but starve the heart of blood slowly through the gradual, progressive narrowing of the artery. This process is likely to cause angina, a severe, crushing chest pain often described as 'a band round the chest'. The pain usually occurs when patients start to exercise or engage in physically demanding activities. This is because the constriction in the artery prevents blood flowing through at the rate needed to fuel the extra work of the heart. Plaques rich in lipids, on the other hand, are more likely to rupture and so suddenly starve the heart of blood. Usually this happens when blood flowing through the artery ruptures a very thin fibrous cap, exposing the lipid pool below. The exposed lipid then stimulates the formation of a blood clot, which grows until it completely blocks the artery. The whole process may take only a few minutes and the end result is a coronary thrombosis and usually severe, irrepressible pain. If the blockage is not cleared, the heart muscles supplied by the affected artery will die. Death of just part of the heart is known as a myocardial infarction.

When a person suffers atherosclerosis or a heart attack, cholesterol is always there en masse and in the thick of the action – and several lines of evidence suggest that high cholesterol level is actually the orchestrator of the illness. In a huge geographic study involving more than 21,000 autopsies in fourteen countries, national levels of atherosclerosis correlated with national consumption of fat. More directly, a study in Framingham, Massachusetts, in the United States, tested the blood cholesterol levels of volunteers several times over a period of several years. After death, their coronary vessels were examined. Of the various factors considered, blood cholesterol level showed the best correlation with degree of atherosclerosis.

For many scientists, though, the most convincing evidence comes from research on the genetic illness familial hypercholesterolemia (FH). This is a rare condition afflicting only one person in every 350 and is due to a defect in the specific receptor for light lipoprotein carriers. With the unloading bays closed, the carriers cannot deliver their loads of cholesterol to cells. The result is an extremely high level of 'bad' cholesterol in the blood of individuals with the disease – and this, in turn, has been suggested as the direct cause of the vascular changes that are also characteristic of the disease. As usual, everybody has two copies of the receptor gene – one from their mother and one from their father. People who inherited a faulty receptor gene from only one of their parents tend to have two to three times the normal level of cholesterol in their blood. They suffer heart disease but can be treated with cholesterol-lowering drugs. When both receptor genes are faulty, cholesterol levels are between five and ten times as high as in normal people. With this severe form of the illness, heart attacks before the age of twenty are common. Cholesterol-lowering drugs do little good, perhaps because blood cholesterol levels are so high that even a substantial reduction fails to bring it within normal range.

The conclusion seems clear. As the National Research Council in Washington put it in their publication *Diet and Health*: 'bad' cholesterol is 'centrally and causally important in the pathogenetic chain' leading to atherosclerosis.

The Victims

More people fall victim to coronary heart disease in industrially developed nations than to any other illness. About 6 million Americans suffer from the affliction to some degree and each year about 1.5 million have a heart attack of which about one-third die; a fifth of those who have attacks receive no warning. According to the *Compendium of Health Statistics*, though, top of the table for heart-disease deaths among fourteen developed nations is not the United States but the British Isles. Compared with American men with a death rate per year of just fewer than 200 per 100,000, Scottish men, for example, have a death rate of a staggering 320 per 100,000. Death rates from heart disease in Britain are more than twice those in comparable European countries with little sign of the gap closing. For example, 125 per 100,000 men die of heart disease per year in the Netherlands, seventy-five in France and, further afield, only twenty in Japan. English and American women share the same level of risk at sixty per 100,000 compared to, say, Italian women at twenty-five per 100,000 and French women at fifteen.

Clearly, the figures show that men are more at risk than women, but the threat to the latter should not be minimized. In Britain, 18 per cent of women die of heart attacks (compared with 25 per cent of men). This means that – at around 76,000 deaths a year – six times as many women die from heart disease in Britain as from breast cancer. Moreover, once a woman suffers her first attack she is 70 per cent more likely to die from it than a man.

Cholesterol – and diet's – links to these deaths have been demonstrated often. The first indication came in 1953 with the publication of a perfect curvilinear correlation between the mortality from coronary heart disease and the consumption of fat – all dietary fat – in six countries. Later – 1970 – the study was refined and extended to seven countries; it then showed that a nation's heart mortality was linked specifically to the intake of animal fat. Within Europe, for example, northerners eat the most saturated animal fats – meats and dairy products – whereas southerners have a diet rich in fruit, vegetables and fish and low in fat. Even within China, where heart attacks are rare and where levels of blood cholesterol are generally low, regional variation still shows a correlation between the two.

More direct correlations have also been reported. Two large clinical trials in the United States, surveying as many as 360,000 people, concluded that a diet high in saturated fat massively increases the risk of developing heart disease. Also in the United States, coronary mortality declined after the 1960s hand in hand with a decline in the consumption of animal fat. In Japan, the diet is lean and heart attacks rare. But when a person migrates from Japan to the United States, he or she acquires almost the same risk of dying from a heart attack as the American hosts.

These dangers of 'bad' cholesterol are almost poignantly highlighted by other studies that show the heart-protecting properties of 'good' cholesterol. In one investigation, more than 7,000 middle-aged men were recruited from twenty-four British towns. First, their blood lipid levels were measured; then their medical development was followed for about four years. In all, 193 of the men suffered a heart attack – and, on average, these were the men with lower levels of 'good' cholesterol when the study began. The mean difference was small – 2.7 mg/dl, or about 6 per cent – but it was statistically significant.

Case for the Defence

So, a diet rich in saturated fats raises cholesterol levels, which in turn blocks arteries and triggers heart attacks. The case against cholesterol seems open-and-shut. However, there is a defence – because the prosecution's case is based on ambiguous science. Branding cholesterol as 'bad' is one way to interpret the data but it is not the only way.

The Diet Affair: Tiny Links, Huge Anomalies

The experiments described in the case for the prosecution show that cholesterol levels can be influenced by diet. Great dedication, much money, and powerful statistics were needed to obtain the evidence – but the proof is there. Such proof, though, is not the end of the argument. Evidence of a link could only be obtained in the controlled conditions of a laboratory or clinical trial. The association observed between diet and cholesterol was small: the best reduction in blood cholesterol level that can be achieved by dietary control in adults is about 4 per cent. If cholesterol is really a killer, this is equivalent, perhaps, to adding three months to life expectancy. Larger effects can be obtained by using food specially designed to interfere chemically with cholesterol uptake, but not by changing simple diets. Because of the small effect, studies of people in the real world find it difficult to show any link between diet and blood cholesterol. For example, the influential Framingham study of cholesterol and atherosclerosis in the early 1950s also included some dietary analysis. Almost 1,000 individuals were questioned in detail about their eating habits. No connection was found between diet and blood cholesterol level, though for some reason the results were never published.

To be fair, though, real-world dietary studies are difficult. After all, how can a researcher measure exactly what a person is eating in their everyday life? All most studies do is simply ask people to record – and sometimes weigh – what they eat over a period of 24 hours, then measure their blood cholesterol levels. With such a small window into people's eating habits, perhaps it's not surprising that no consistent association can be found. There is no clear difference in diet between people with high and low blood cholesterol levels, in either adults or children. People who eat lots of animal fat in the 24 hours recorded have no more cholesterol in their blood than people who eat very little. However, it cannot be just the brevity of the time-window that is critical. Even studies that have asked people to meticulously record and weigh what they eat over two weeks, at different times of year, have found no correlation. Despite enormous variation in what subjects eat – consumption of animal fat can vary from person to person from about 10 g/day to over 300 g/day in such studies – a convincing link with blood cholesterol can still not be demonstrated. Extremely low cholesterol values are seen in both those who eat little animal fat and those who eat the most; high cholesterol values, too, are seen at all levels of animal fat intake.

Such results hardly seem grounds for initiatives to change a nation's coronary health by changing its diet. The American National Cholesterol Education Program continues to urge all Americans to reduce their intake of animal fat to 10 per cent of their total calories. Yet in many of the real-world studies, a substantial proportion of the volunteers – up to around 15 per cent in some – already ate that little animal fat but their cholesterol levels were no lower than those who ate much more. This probably also explains why there is no clear sign of cholesterol levels falling despite heightened public awareness of cholesterol and a real change in eating habits. In Britain, for example, blood cholesterol levels do not seem to have fallen since the first survey in 1974. However, the absence of a comprehensive system for monitoring blood cholesterol levels in Britain might make it difficult to pick up changes for specific age groups.

Alongside the subtle experimental evidence that diet can influence cholesterol levels in the way expected – and alongside the unimpressive real-world studies – there are some huge natural anomalies that need explaining. For example, some cultures regularly eat enormous amounts of animal fat without having high levels of blood cholesterol; on the contrary, levels may be very low. An average African Samburu, for example, eats about a pound of meat and drinks almost two gallons of raw milk every day for most of the year. Moreover, the milk from the local Zebu cattle contains much more fat than cow's milk. On average, Samburus consume more than twice the amount of animal fat than the average American, and yet their cholesterol is much lower, about 170 mg/dl as opposed to about 220 mg/dl. Similarly, a Kenyan Masai drinks half a gallon of Zebu milk each day and on special occasions literally parties on meat – several pounds per person is commonplace. Yet the average cholesterol level of Masai tribesmen at 135 mg/dl is among the lowest ever measured in the world, about 60 per cent of the value of the average affluent American male.

Undoubtedly, part of the explanation for such low cholesterol levels in the face of such high-fat diets is genetic; the Masai body might adjust the manufacture and breakdown of cholesterol differently from the average American. This does not mean, though, that Masai cholesterol levels cannot change. For example, a study of a group who moved from the African bush to the Nairobi metropolis found that blood cholesterol level rose by 25 per cent – despite the new diet inevitably including less animal fat than the old. Something about these Masais' new lifestyle had triggered an increase in blood cholesterol level – but it wasn't their change in diet.

It isn't difficult to reconcile all these conflicting pieces of information – but to do so exonerates diet from complicity and throws cholesterol's role in the body into a new perspective. The body 'knows' how much cholesterol needs to be circulating in its blood. After all, cholesterol is a useful chemical that does many vital jobs – from waterproofing cells to manufacturing hormones and bile – and a certain amount is always needed. How much is needed will have been fixed by evolution in its usual exquisite way as our ancestors meandered through ancient environments and lifestyles. Like most evolved attributes – as will be discussed for depression – the optimum level will differ from person to person according to differences in, say, body-type, lifestyle and behaviour. For any given person in any given situation there will be an optimum – but individually different – blood cholesterol level that the body should strive to maintain, buffering itself from day-to-day variation in diet. It can do this because it manufactures most of its own cholesterol – and compensation mechanisms exist that keep cholesterol levels steady. If the amount of cholesterol in the diet goes down, the body increases production – and vice versa. Around twenty genes are already known to be involved in the cholesterol pathway, including some involved in regulation and individual differences. For example, a brain chemical known as Neuropeptide Y (NPY) plays a role in regulating blood cholesterol levels – and people with one form of the gene that produces NPY have significantly higher blood cholesterol levels than people with other forms.

It isn't surprising, then, that even controlled experiments find only a small influence of diet on cholesterol level – or that in the real world it is difficult to find any convincing relationship between the two. Nor is it surprising that when lifestyle changes hugely – such as when swapping the African bush for a metropolis – huge changes in cholesterol level can occur that are in the direction opposite to that expected from change in diet. The implication is that even when blood cholesterol level declines in the face of a low animal-fat diet, it is not because the body is being starved of cholesterol – or the means of manufacture. The changes are too subtle. Much more likely is that the change in diet is associated with other changes – subtle alterations in lifestyle, behaviour or physiology – that require the body to reduce its average blood cholesterol level.

Maybe, then, diet is innocent of all complicity with cholesterol. When scientists see a subtle change in blood cholesterol level in the direction they expect following a change of diet, perhaps they are seeing only an indirect response. When in the real world they have difficulty seeing any change, perhaps it is because more important factors than diet are at work. And when they see a huge change in a direction they do not expect, perhaps they are seeing an even greater response of those same factors.

But does any of this matter? If a change of diet or a lowering of blood cholesterol level still leads to fewer strangled hearts and deaths, does it matter how it happens? Even if the evidence is unconvincing, perhaps national campaigns to persecute high-fat diets and cholesterol could still be justified.

Does Cholesterol Really Strangle Hearts?

The evidence that diets rich in saturated fats or high cholesterol levels increases the degree of atherosclerosis is weak. This includes the findings of the fourteen-country study involving more than 21,000 autopsies and the Framingham study that both seemed so convincing in the hands of the prosecution.

The best correlation in the fourteen-country study was between atherosclerosis and the total intake of fat. Intake of animal fat alone was not a significant factor. This implies, rather disquietingly, that vegetable fat is as, if not more, dangerous – which makes no sense in terms of the case against cholesterol. However, the correlation was probably never robust in the first place. For some reason, Japan was excluded from the study. If included, the correlation would probably have disappeared. The Japanese at that time (1968) had a low intake – about one-third that of Americans – of fats and animal fat. They also had lower blood cholesterol levels than in the United States – 170 mg/dl as compared with 220 mg/dl. Yet the degree of atherosclerosis in the two countries was similar. In fact, a later study of the arteries of the brain in 1,408 Japanese and more than 5,000 Americans showed for all age groups that the Japanese were marginally more arteriosclerotic than Americans.

Even the influential Framingham study is inconclusive. It did show that blood cholesterol level was more associated with atherosclerosis than, for example, age, weight and blood pressure. Although the correlation was weak, it was statistically significant. There is a suspicion, though, that the association was an artefact of non-random autopsies leading to an over-representation of subjects with the genetic disease FH. These people have high levels of both blood cholesterol and atherosclerosis but are not representative of the population at large. Inclusion of just one or two people with this condition might have been enough to produce a weak correlation of the order recorded.

Random selection of subjects is essential in such studies – especially if people with FH cannot be identified and excluded. When survey design is solid, any link between high levels of blood cholesterol and atherosclerosis seems to disappear. As long ago as 1936, a New York University study of people who had died violently demonstrated that there was no link between cholesterol and arterial disease for any age group. Violent death is not a random selection process, as we shall see when discussing depression, and cholesterol values in dead bodies may not mirror values when alive – but the survey was more solidly designed than most. A Canadian study avoided the problem of posthumously measured cholesterol levels by monitoring about 800 World War II veterans, taking blood samples over a number of years. Veterans that died between the ages of sixty and seventy showed no correlation between degree of atherosclerosis and blood cholesterol level; those with low cholesterol levels over the years were just as arteriosclerotic when they died as those with high cholesterol levels. Similar conclusions have been reached from studies in India, Poland, Guatemala, and the United States.

If there is a link between blood cholesterol level and degree of atherosclerosis, it isn't obvious.

Heart of the Matter

When it comes to the crunch, though, does it really matter whether a diet rich in animal fats raises blood cholesterol levels or even whether high blood cholesterol level hardens arteries? The crucial factor is whether diet and/or cholesterol levels are associated with the risk of death from coronary heart disease.

The early six- and seven-country analyses that seemed to show a link between national coronary mortality and national consumption of fat did nothing of the sort. They were correlations, not proof of cause and effect. The choice of which countries to include was selective and the statistics made no allowance for whether the same association was evident within countries as between countries. In Finland and Greece for example, heart mortality showed great variation between regions despite broadly similar diets and other risk factors. More recent and less eclectic studies either fail to find any association or find it to be very weak. One of the largest was the 'Monica study', funded by the World Health Organization, which assessed twenty-one countries over ten years yet found no statistical connection between reduction in coronary heart disease and changes in a wide range of factors, including cholesterol levels. The final conclusion was that the results gave more insight into the difficulties of running trials over huge geographical areas than they did into the role of cholesterol and other risk factors.

So many factors vary from country to country that it is very dangerous to deduce cause and effect from any international correlation, especially a weak one. More direct studies are needed. For example, if dietary fats – particularly animal fats – are really the villains portrayed, we should at least expect to see this reflected in the diets of people who have had a heart attack. Yet the evidence is far from persuasive. In only one study out of thirteen did patients with coronary heart disease eat significantly more saturated fatty acids than healthy individuals of the same age and sex; a rate hardly better than pure chance.

In a few countries, the heart-attack rate over the decades has

shown ups and downs parallel to those of dietary saturated fats –
but in others it has not. In some countries, fat consumption has
changed but heart mortality has not – or vice versa – and in a
few countries they have even changed in opposite directions. For
example, in Switzerland, the intake of animal fat increased by 20
per cent after World War II, yet coronary mortality decreased. In
England, the intake of animal fat was relatively stable between
1930 and 1970, yet the number of heart attacks increased tenfold.
Coronary mortality also increased tenfold in the United States –
between 1930 and 1960 – while the consumption of animal fat
was actually declining. Even the decline in American coronary
mortality since the 1960s – which does coincide with a decline in
the consumption of animal fat – may not be all it seems. Data
from Framingham show that the post-60s decrease in coronary
mortality was actually balanced by an increase in the number of
non-fatal heart attacks. Maybe, therefore, the impressive American decline in mortality has had more to do with improving treatment than improving diet.

The example of Japanese migrants to the United States is also
not as convincing as it first seems. True, Japanese migrants to the
United States subsequently died from heart attacks almost as
often as their American hosts; much more than if they'd stayed in
Japan. As with the Masai in Nairobi, however, change in lifestyle
seemed more important than change in diet. Those who lived
within Japanese communities in the United States, even if they ate
American food, experienced much less of an increase in heart
attacks than those who did the converse, embracing the American
way of life but eating Japanese food.

The evidence that a 'bad' diet is detrimental to coronary health
therefore hovers between weak and non-existent. What about the
evidence that tests directly whether cholesterol levels influence
coronary health: that 'good' cholesterol protects and 'bad' cholesterol destroys? The British study of 'good' cholesterol and the
Framingham study of 'bad' were not representative; simply the
ones that best fitted the advice being given by various national
heart associations. If human coronary health really is sensitive to

'good' and 'bad' cholesterol, it is very reluctant to yield the fact to research. Most studies find no link. The postulated heart-protecting value of 'good' cholesterol has been studied on innumerable occasions – because even in total the evidence is no better than ambivalent. The search for conclusive proof still continues. If 'good' cholesterol does protect, its role must be minor: there are people with almost no 'good' cholesterol who don't get heart disease. Even the links that are found are not always as expected. Sometimes 'good' cholesterol emerges as the more important, sometimes 'bad'. Sometimes, inexplicably, total cholesterol level is the more important, as if 'good' and 'bad' are both acting in the same direction, rather than against each other.

An illustration of the brittleness of even the better-designed studies is the British investigation of 7,000 men from twenty-four towns, which set out to explore the protective role of 'good' cholesterol. When, as we have already seen, the researchers simply compared the cholesterol levels of the 193 men who died of heart attacks with those who didn't, they found a significant difference of 6 per cent in the direction expected. However, when they took other risk factors into account, described later, the protective role of the 'good' cholesterol disappeared within the statistical maze. A few years later, when the number of heart attacks in the study group had risen to 443, the influence of 'good' cholesterol reappeared, only to be inexplicably dwarfed by the influence of total cholesterol, which in effect means that there was no 'good' cholesterol, only 'bad'. This is confusing enough, but who knows what would have happened to the rickety statistical edifice had the researchers been able to add yet further risk factors to their analysis?

Guilty: But Are There Grounds for Appeal?

Having surveyed all the evidence, the scientific jury of the 1970s reached a clear majority verdict: fatty diets and 'bad' cholesterol were guilty of all charges. As a result, every effort has since been made to protect Western societies from these greasy conspirators. Affluent countries for which coronary heart disease is a major killer have zealously educated people about the evils of animal fat and the virtues of polyunsaturated vegetable fats. In adults, a cholesterol level of less than 200 mg/dl of blood has been deemed desirable. More precisely, the level of 'bad' cholesterol should be no more than 160 mg/dl of blood and that of 'good' no less than 35 mg/dl.

Most of the developed nations have had at least partial success in their attempt to change their people's diets. In Britain, for example, although total fat consumption has stayed relatively steady at around 40 per cent, the polyunsaturated/saturated (P/S) fat ratio has been gradually rising – for example, from 0.22 in 1977 to 0.37 in 1987. In the United States, average calorific intake from saturated fats is still above the 10 per cent target at 13 per cent and, although the average woman has an average cholesterol intake at about the target of 300 mg/day, the average man is still far away at 435 mg/day. Although there may still be some way to go to meet dietary targets, the United States, Canada, Western Europe and Australia all claim that their campaigns are succeeding as far as coronary mortality is concerned. Since the mid-1960s, death rates have plummeted by 60 per cent in North America, and by up to 40 per cent in Western Europe. They have also fallen steeply in Australia. Only in Eastern Europe are the death rates from heart disease rising.

At first sight, then, it would seem that the scientific jury was right and that their advice to policy makers was justified. Diet and cholesterol were the culprits and the campaigns are working. However, we cannot unquestioningly accept that the observed declines are due to the national dietary campaigns. Many factors have changed over the same time period, including the timing of diagnosis and quality of treatment. And there are anomalies. For example, age-adjusted death rates from coronary heart disease have also fallen in Japan. Yet – with low levels of heart disease from the beginning – the Japanese saw little reason to influence the nation's dietary evolution. Consumption of dairy products and red meat actually increased in Japan during the 1970s and 1980s. So, too, did cholesterol levels until in some areas they were on a par with those in Britain. But Japanese rates of heart disease remain low and British rates remain high. Despite the improvement in P/S ratios in the British diet, blood cholesterol levels do not seem to have responded. Neither have rates of heart disease – at least not as impressively as elsewhere. Estimates for men are for a modest decline of between 1 and 15 per cent; for women they range from a decline of 8 per cent to an increase of 12 per cent. Much depends on age group.

Social group is also important – and throws up yet another anomaly. Plump, senior civil servants with a diet of rich food are less likely to have heart problems than their lean, diet-conscious subordinates. Since the late 1960s, the 'Whitehall study' has been monitoring the health of 17,000 civil servants. They found that the lower the grade of civil servant, the higher the incidence of coronary heart disease. This is part of a more general disappearance of heart disease as a killer of upper- and middle-class professional men under sixty-five. Deaths – which topped 200 per 100,000 in the 1970s – have now halved for this group of men and are continuing to decline.

Such anomalies make it difficult to read anything into gross national statistics for the last thirty years. More direct evidence is needed if national campaigns to control diet and cholesterol are to be credited with improving coronary health. And there has been no shortage of attempts to obtain that evidence. More than forty trials – involving well over 100,000 people – have now tested whether the deliberate lowering of cholesterol levels – by diet or drugs – prevents heart attacks. In some of the trials the number of fatal heart attacks was lowered a little, but in others they actually increased. A summary concluded that on average whereas slightly fewer people – 2.8 per cent compared with 3.1 per cent – suffered non-fatal heart attacks after lowering their cholesterol, the number who suffered fatal heart attacks stayed the same at 2.9 per cent. In fact, the number of deaths from all causes seemed to go up – from 5.8 per cent to 6.1 per cent –

suggesting that high cholesterol may even have a protective role against some life-threatening factors.

Undeterred by the mess, various statistically minded biologists have delved into the data seeking at least some crumb of support for the original guilty verdict. Meta-analyses have been performed, combining the data from many tests. This is an indictment by itself: if the evidence were strong such statistical jiggling would be unnecessary. But even after massive number crunching, nothing is clear. Some meta-analysts claim that the deliberate lowering of cholesterol does reduce heart attacks. Critics, though, say that this is only true if one or more trials with a negative outcome are omitted – or if the statisticians are selective over which categories of patient they include. For most of us, the truth stays hidden among the mass of confusing data. But it is difficult to avoid the suspicion that if, after so many trials, the effect of lowering cholesterol isn't obvious to all, then there is no effect. Perhaps the original verdict was wrong and high-fat diets and high cholesterol levels are innocent after all.

In the last few years, though, there has been a new development. In contrast to dietary change and in contrast to trials of earlier cholesterol-lowering drugs, trials of some of the more recent drugs – the so-called statins such as Zocord and Pravachol – do produce convincing results. Statins lower cholesterol levels – and reduce coronary mortality. Does this mean that 'bad' cholesterol is guilty after all, even if a diet rich in saturated fats is not? Not necessarily – because the success of statins may not be due to their lowering of 'bad' blood cholesterol levels.

Statins reduce cholesterol levels because they inhibit the body's production of mevalonate, a precursor of cholesterol. But mevalonate is a precursor of other substances too. One such substance, for example, makes blood platelets less inclined to clot and smooth muscle cells less active – cutting down on their growth and migration inside artery walls. Statins could thus prevent cardiovascular disease by at least two mechanisms, neither of which involves cholesterol.

Quite possibly, therefore, the lowering of cholesterol level and the reduction in heart disease are two independent effects of statin treatment, and some aspects of the trials support this notion. Statins reduce coronary mortality for just about everybody, even those for whom cholesterol level would not normally be considered a risk factor. As just one example, statins reduce the risk of coronary heart disease even for people with a low cholesterol level for whom conventional wisdom would say further lowering would not reduce the risk. The only reasonable explanation for this and various other effects is that statins do not reduce the risk of heart disease by lowering cholesterol but by one of their other – probably numerous – effects.

The possibility that high-fat diets and 'bad' cholesterol were never guilty as claimed is still very real. There are certainly grounds for an appeal and a fresh look at all the evidence.

But If Cholesterol Is Innocent, Who Is Guilty?

Over twenty years after the scientific establishment returned its guilty verdict, it seems more than possible that there has been a miscarriage of scientific justice. But if cholesterol is innocent of causing heart attacks, where lies the guilt?

The Conspiracy Theory

Among all the arguments and counter-arguments, there is still one central fact that nobody disputes and that demands explanation. People with high levels of cholesterol in their blood are more likely to develop coronary heart disease than people with 'normal' or low levels. How can this fact be squared with cholesterol's innocence? The statin trials just described provide a clue. Perhaps factors exist that not only increase the risk of coronary illness but at the same time raise the level of cholesterol in the blood. If so, these factors could be the real guilty parties, responsible not only for causing death but also for conspiring to make cholesterol take the blame. This is a common enough deception in real crime; and it is probably a common enough deception in biological systems.

There is already a long list of 'lifestyle' factors that have such a dual influence: obesity, diabetes, smoking, a sedentary lifestyle, poverty, unemployment, low birth weight, stress, high blood pressure, unrest and sleep disturbance are all associated with both raised cholesterol levels and increased risk of coronary heart disease. As far as the link with cholesterol is concerned, all influence total blood cholesterol and some – such as obesity, smoking, high blood pressure and activity levels – also influence the ratio of 'good' to 'bad' cholesterol. For example, sedentary people have, on average, 6 per cent more total cholesterol than 'walkers' (those who take brisk daily walks). The excess is all 'bad' cholesterol – because 'good' cholesterol is actually reduced. Sedentary people have 27 per cent less 'good' cholesterol than walkers and 41 per cent less than distance runners. And as far as the link with heart disease is concerned, the evidence against most of the potential conspirators is far more convincing than the evidence against cholesterol has ever been.

Smoking, for example, causes almost one in five coronary heart disease deaths. Under fifty years of age the risk of a heart attack for a smoker is up to ten times that of a non-smoker the same age. When a smoker in his thirties or forties has a heart attack, there's an 80 per cent chance that it was caused by tobacco. People who smoke twenty cigarettes a day face a 75 to 80 per cent higher risk of developing heart disease than non-smokers. And the more a person smokes, the greater the risk. If a person stops smoking, though, the risk halves after only one year. Even passive smokers are at risk. Although they breathe in only about 1 per cent of the amount of smoke inhaled by direct smokers, some studies claim that the increase in risk of a heart attack may be as high as 20 per cent.

Weight has a double link with heart disease – in its own right and via diabetes. For example, middle-aged women who have put on more than 10 kg in weight since they were eighteen years old are two to three times more likely to die of heart disease than those who have stayed at about the same weight. If they put on 20 kg they are seven times more likely to die prematurely. But being overweight also increases the risk of developing diabetes, which in turn increases the risk of heart disease still further.

Poverty increases the risk of heart disease in a variety of interrelated ways. Unemployment, stress, high blood pressure, general unrest and sleep disturbance are probably all involved. So, too, is malnutrition during gestation and early infancy. In Britain, for example, there are heart-disease hot spots. In northwest England, people are twice as likely to die of heart disease as people in, say, London. The areas where death rates from heart disease are high today match those where death rates in newborn babies were high sixty or more years ago; factors that threaten people as babies seem to raise their risk of heart disease in later life. The main links seem to be low birth weight and blood pressure. Underweight babies are more likely to grow up to have high blood pressure. Moreover, adults who were very small as babies and adults with heart disease both tend to have a thick wall in the left ventricle of their hearts. Quite possibly the undernourished foetus adjusts its physiology and anatomy to maximize its chances of short-term survival at the expense of long term.

The list of potential conspirators to link cholesterol with heart disease is almost certainly incomplete. There will be more as yet unknown. For example, people from the Indian subcontinent have an unusually high risk of developing heart disease – and they also have high blood cholesterol levels. Some factor is at work that is a potential conspirator. It could partly be that South Asians have a genetic predisposition to develop diabetes. But it could also be that they have narrower coronary arteries than white Europeans, their diameter averaging almost 10 per cent less. A blood clot of a certain size in the coronary arteries of two people is more likely to cause a blockage – and a heart attack – in the one with a narrower artery. Perhaps even beyond Asia people with narrow arteries often have a raised cholesterol level, adding yet another factor to the list of potential conspirators.

Not all factors that increase coronary disease, though, fit the conspiratorial profile – because they are associated with low cholesterol levels, not high. For example, as we shall see in the next chapter, people who suffer from severe clinical depression are up to four times as likely to die from heart disease as people who are not depressed. Yet depression tends to be associated with low cholesterol levels.

Depression apart, all of the major risk factors for heart disease simultaneously increase cholesterol levels. They are potential conspirators against cholesterol but not certain – proof is needed. After all, how can we tell what is causing what? Do these various factors increase the risk of heart disease because they increase cholesterol, employing the lipid to do their dirty work? Or do they increase the risk of heart disease independently of their boost to cholesterol, thus artificially making the innocent lipid look guilty? One way to tackle such questions is statistical: to employ multiple-factor techniques designed to tease apart the relative importance of a range of potentially important factors. An example is the British investigation that set out to explore the protective role of 'good' cholesterol. When, as we have already seen, the researchers controlled statistically for other factors, the protective role of the 'good' cholesterol disappeared within the statistical maze. In this case, these 'other factors' were age, blood pressure, body weight, and cigarette smoking. Statistically, it seemed that men suffered heart attacks because they were older, fatter, had higher blood pressure and smoked more, not because they had 'unhealthy' cholesterol profiles.

A major problem with such analyses, though, is that a factor may look potent in the statistical company of one set of risk factors, insignificant among another set – then may look potent again among a third. It is difficult to know which analysis to believe and when to stop; conclusions are always fragile, just waiting to be overturned by the addition of a new factor. Who knows what would have happened to the evidence for or against 'good' cholesterol had the British researchers been able to add yet further risk factors, such as stress, exercise or infection levels, to their analysis? All conclusions could change if some previously unsuspected factor entered the statistical arena.

Disease and Inflammation: The Missing Factor?

No scientist can ever know whether he or she has discovered all the relevant risk factors; the important factor could still be awaiting discovery. The best that can be done is to try to judge how much of the explanation is still missing. For example, the decline in coronary heart disease over the past few decades in so many developed countries cannot easily be explained by amelioration of all the known risk factors. True, the proportion of men who smoke in the West has dropped from half to about a third – and average blood pressure has also fallen. Malnutrition during gestation and infancy is also less common. However, obesity and diabetes have both become more common. If we include Japan in the equation as well – where most risk factors have increased yet coronary disease has fallen from low to lower – the difficulty becomes even more apparent. Conservative estimates suggest that all the lifestyle improvements together account for no more than half of the post-60s decline in coronary death rates seen in many countries. So how do we explain the other half?

One possibility – with a few avid supporters and many vitriolic opponents – is that the starting point for coronary heart disease is inflammation of the arteries caused by infection. It has long been recognized that inflammation of the artery wall is an early step in atherosclerosis, but as described earlier most scientists blame that initial irritation on damaged cholesterol carriers. It is possible, though, that inflammation comes first – and cholesterol deposition comes shortly afterwards. There is some evidence for this. If inflammation of the arterial lining of rabbits is disrupted, plaque does not form, even if the animals have high levels of cholesterol in their bloodstream.

Arterial inflammation could, of course, be triggered in many ways – for example, chemicals in tobacco smoke are arterial irritants when in the bloodstream – but among the most potent would be microbial infection. French research has shown that mice exposed to air containing microbes suffer damage to their circulation system. The damage – in the form of plaques – looks like atherosclerosis. There is also a genetic element: the risk of plaque formation was greater if a naturally occurring protein – called IL-10 – was lacking. Mice genetically unable to produce IL-10 had blood vessels thirty times more clogged than other mice. The walls of the vessels were also four times more likely to rupture. Some humans – up to 10 per cent – also produce low levels of IL-10 and could be especially vulnerable to heart disease from microbial infection.

One microbe in particular has aroused interest: the bacterium *Chlamydia pneumoniae* – often called TWAR after the original laboratory strain. It is spread by coughs and sneezes – and by the time a person is twenty, there's a fifty-fifty chance of carrying the infection. Migration from the lungs to the coronary arteries could be by hitching a ride on macrophages. During the 1990s, the fact slowly emerged that people with coronary heart disease often have high levels of antibodies to TWAR, suggesting that they have been persistently or repeatedly infected. Furthermore, the bacterium's DNA and proteins often show up in plaques from arterial walls – and on occasion even the live bacterium has been found. Signs of the bacterium have appeared in 60 to 79 per cent of atherosclerotic arteries in studies in Finland, Italy, Britain, Japan and the United States – but have rarely been found in healthy arteries. A few other studies, though, have found no sign of the bacterium in diseased arteries and suggest that the positive results elsewhere are the result of laboratory contamination.

The rise of heart disease in the early twentieth century and its recent rapid decline has often been termed an epidemic – because it follows the classic pattern of an infectious disease, albeit over an unusually long period of time. Maybe heart disease has been an epidemic in a more real sense than anybody ever supposed. A bacterial cause could explain why death rates from heart disease are falling more rapidly than expected if they were simply mirroring the decline in classical risk factors. Broad-spectrum antibiotics

– such as tetracycline – which kill chlamydia, were introduced at about the same time death rates started to fall.

Of course, we have to be careful not to make the same mistake as for cholesterol – brand chlamydia a killer when it is just an innocent bystander, present in the arteries but doing no damage. There is more reason, though, for expecting chlamydia to be guilty. Whereas cholesterol occurs naturally in human arteries, chlamydia does not – and C. *pneumoniae* has family members already known to wreak havoc with human body plumbing. A close relative is one of the main causes of infertility in humans, blocking oviducts in women and vasa deferentia in men. It is quite feasible that microbial infection could fill the void for the missing 50 per cent of explanation for human heart disease.

The Scale of Injustice: Is Cholesterol Actually a Healer?

There is a strange and final dimension to the cholesterol affair. It comes when we ask ourselves why the lipid is always present when hearts are strangled. And why do most factors that increase the risk of heart disease also increase cholesterol levels? If cholesterol is the victim of conspiracy, it has almost gone out of its way to bring the injustice on itself. There is an alternative but it involves turning the whole of the cholesterol saga on its head, escalating the scale of any injustice. Perhaps the lipid's omnipresence is because one of its jobs is actually to heal damaged arteries.

Cholesterol is an important chemical in the human body. It is vital for the healthy membrane structure of cells, for fertility and for embryonic development; pregnant rats given drugs that block the manufacture of cholesterol produce deformed foetuses. It is also involved in the production of hormones and bile – and probably in many more biochemical pathways still unknown. The body regulates cholesterol levels carefully, adjusting production and regulating traffic according to its needs. On a day-to-day basis, the body keeps blood cholesterol level relatively constant despite short-term changes in diet, but it can also make huge adjustments when necessary. When people smoke, grow fat, become stressed, become infected, etc., their bodies' cholesterol needs must change. The fact that most factors associated with coronary heart disease lead to an increase in cholesterol production could mean that when the illness looms, the need is for more cholesterol, not less. Why should this be?

Smoking irritates arterial linings. So, too, do microbes like chlamydia. Stress, high blood pressure, obesity and diabetes may also either cause arterial irritation directly or reduce resistance to factors – such as microbes – that actually cause inflammation. Suppose that the deposition of cholesterol on arterial walls is actually the body's way of defending the arteries against such irritation and inflammation. Cholesterol production and traffic would need to increase – particularly if the fatty streaks are not just formed and then forgotten but are also maintained, old cholesterol being removed and replaced with new. Some streaks may be temporary, removed by the heavy lipoprotein carriers once their job is done and the arterial wall healed. It is already clear that fatty streaks do not have to be dangerous. Japanese people show as much, if not more, atherosclerosis as Americans yet suffer far fewer heart attacks. In our evolutionary past, the mechanism clearly worked: heart disease was a twentieth-century epidemic of the Western world. Perhaps it is only when arterial irritation is long-term or intense and caused by the artefacts of the Western lifestyle that cholesterol cannot cope with the situation; the need for macrophages and platelets to shore up ever more diseased tissue eventually leads to a heart attack.

As cholesterol is such an important chemical – maybe even more important than we realize – we probably mess with its activity in our bodies at our peril. The genetic disease FH provides a graphic illustration of what happens when the normally exquisite cholesterol control system goes awry. Maybe the disease does emphasize the dangers of excess cholesterol as traditionally claimed, but it could equally well show the difficulty of healing inflamed arteries when the cholesterol-deposit system is not working properly. Cholesterol-lowering drugs seem to help such patients – as long as they have only one copy of the FH gene. The help, though, may stem from mechanisms other than lowering cholesterol, as we saw with statins. For the vast majority of people who do not have FH, the deliberate lowering of cholesterol levels by drugs may have unforeseen and detrimental effects. Trials of such drugs have already reported that the drug takers were more likely to die from cancer, accidents, suicide and murder, suggesting that either the drugs – or cholesterol – were influencing behaviour as well as physiology. This is not so surprising for a chemical as involved in hormone production as cholesterol. The beneficial effect of the drugs on heart-attack rate – for reasons that may have been independent of cholesterol levels – were balanced by a downside: the drugs did not always reduce overall mortality rates.

In the study of heart disease, scientists – like many a jury – have been so swayed by the suspect's omnipresence at the scene of the crime that guilt seemed obvious. Obtaining proof then became something of an obsession, and as frail study followed frail study, efforts were simply redoubled in an attempt to pin the quarry down. The science was weak at the outset – mainly simple correlations, ultra-vulnerable to artefact – and became no more robust as time went on. It became shored up, though, by a multimillion-dollar food industry, plus medical initiatives and advice that would be embarrassing to reverse or retract. Maybe, despite all that, the original verdict was sound and the treatment and advice correct. But maybe the result has been a miscarriage of scientific justice with a trail of innocent victims: a useful molecule in our blood; the producers and manufacturers of animal fat all over the world; and millions of healthy people who have been cajoled into changing their diet to rid them of a chemical that they may even need.

3

Clinical Depression and the House of Cards

Clinical depression is a distressing and disabling illness – and according to the hype it is on the increase and here to stay. The Prozac generation will become the Prozac century, the Prozac Nation will become the Prozac World, and before long humankind will only function on the back of antidepressants. The most pessimistic figures for Western societies state that up to one in three people can expect to have a severe, disabling bout of depression some time in their life. And without antidepressants – perhaps even with antidepressants – up to one in six of these could commit suicide. And that's just in the developed world. In the developing world, according to the World Health Organization, depression will become the number one cause of disability by 2020, just ahead of road-traffic accidents, heart disease and lung disease. Is our mental health really in such a dire and deteriorating state? And if so, why? Is it that increasing numbers of children are having such a miserable, unstable and abused childhood that they cannot help but sink into depression in adult life? Or that increasing numbers of people are screwing up the sensitive chemistry of their thought processes by taking medications or recreational drugs?

Depression is already a huge drain on Western society, its immediate cost the devastating effect on the victims, their families and partners. There is also a wider financial and social cost. In the 1990s, depression was draining American society by around $45 billion each year. More than half of that bill was paid for by businesses in the form of lost productivity and absenteeism.

Treating depression cost $12 billion, and depression-related sui-
cides added $8 billion through the loss of future earnings.

At the same time – the other side of the coin – depression is big
business. Without it, there would probably be fewer mind-
consultants in employment – and pharmacologically, there are
already more than thirty antidepressants on the market, with
sales booming. Few people can have failed to see the hype about
Prozac – the Dog Star among antidepressants – and most know
at least one person who is taking the drug. In recent years,
pharmacists' shelves have filled up with Prozac-type antidepressants and tens of millions of people around the world have
swallowed the drugs on the implicit understanding that not only
do they work but their actions are specific and scientifically
proven. But how much does science really know about clinical
depression, its causes and its treatments? Are we all equally at
risk and is it really on the increase?

Ever Higher Lows?

The number of cases of clinical depression being reported in
Western societies has been increasing since at least the 1970s,
giving the impression of a decline in the mental health of industrial nations. An increase in reported cases, though, does not
necessarily mean that the illness itself is becoming more common.
Maybe the apparent increase is an artefact due to better – or
more cavalier – diagnosis by the medical profession. Or perhaps,
as the social stigma historically attached to mental illness wanes,
increasing numbers of people are consulting their doctors for
advice, people who in the past would either have suffered in
silence or relied on some form of self-help. How can we tell?

We can't – and for two main reasons. First, there has been no
centralized historical system for monitoring depression – in contrast to, say, criminality and drug-use – so there is no reliable,
central yardstick against which modern measurements can be
judged. Secondly, doctors' records – the only place where histori-

cal data might be buried – are unreliable even now. A report in the *British Medical Journal* in 1992 claimed that every year doctors diagnosed 3 per cent of the general population in Britain as suffering from depression. But they misdiagnosed a further 3 per cent who in reality were also suffering from the condition. In addition, many more depressed people felt that they were not ill enough to visit a doctor. The report concluded that – as a best guess – at least half of all cases of depression in Britain go undetected. In the past that proportion might have been even higher. Without knowing, we clearly cannot use doctors' records to investigate whether depression is on the increase.

The only other possibility – and an approach that has been used – is to carry out surveys of populations every few years and either to compare them with past surveys (though the first wasn't done until the 1970s) or to rely on people's self-diagnoses and memories. In principle, the process seems reasonable – and the results are virtually unanimous: a wide range of studies in different countries has concluded that there has been a significant increase in depressive disorders. How robust, though, are these conclusions? Undeniably, most such surveys are carried out thoroughly by researchers well trained in interview techniques and using carefully formulated questionnaires. Their conclusions, though, depend critically on people honestly answering questions about their current mental state and often on their ability to remember past states. They therefore depend on people's attitude having stayed the same over the years – but it hasn't. Since the first studies in the 1970s, public awareness of depression has changed, the stigma attached to mental illness has changed, and even the definition of depression has changed. Any one of these changes could make depression, however it is measured, seem to be increasing when in reality it is not.

Diagnosis

There is no objective – and therefore no reliable – method for diagnosing depressive illness. There is no blood test or brain scan that can say unequivocally 'this person is suffering from clinical depression'. Diagnosis depends instead on either the subjective judgement of a doctor or psychiatrist after interviewing the patient or on some 'score' on a questionnaire.

The subjective diagnosis of clinical depression is difficult for even an experienced clinician. The illness varies considerably from person to person – and even more from culture to culture. Not everybody experiences all aspects of the disease. The most consistent psychological symptoms in Western sufferers are feelings of worthlessness and despair, including thoughts of suicide. Often, also, there are physical symptoms such as stomach complaints, a rapid heartbeat and headaches. Some sufferers may actually look depressed, with the eyes and corners of the mouth turned down, the face somehow frozen into a grieving mask. They may cry frequently, have difficulty forcing themselves to move and even the smallest tasks may seem to require superhuman effort. Nothing gives them pleasure and when they ask for help, nothing satisfies them. Lowered self-esteem and feelings of hopelessness may go hand-in-hand with blaming themselves for their state and the feeling that their families and friends would be better off without them. Frequently they feel that they are losing their minds and suffer delusions and hallucinations. Although depressed patients are often passive and withdrawn, they may also have moments of irritability, anger, hostility and paranoia.

Everybody 'feels low' on occasion, for periods lasting anything from several minutes to perhaps several days. So a major problem in diagnosis is how to separate such transient feelings from real clinical depression. Some ad hoc yardstick has to be applied and, according to current guidelines, a loss of interest or pleasure in nearly all activities does not qualify as clinical depression until it has lasted relentlessly for at least two weeks. Even then, there are gradations from mild depression to major or severe depression, depending on how many of the above symptoms are shown by the patient for that time. In reality, there is probably an unbroken continuum – at least of symptoms – from mild blues to anxiety through to full-blown, life-threatening clinical depression.

Yet diagnosis and treatment requires a line to be drawn somewhere. If, since the 1970s, that line has been drawn progressively further towards milder cases, an increase in reported cases would have occurred without any change in incidence of the illness itself.

Suicidal Yardsticks

One way around the problem that the yardstick for diagnosis has changed over the years might seem to be via the statistics for suicide. The vast majority of suicides – as many as 90 per cent – are committed by people suffering from depression, suggesting that variation in suicide rates could give some indication of the way that the incidence of depression has changed.

Most Western societies show parallel trends in suicide rates but these are up and down with no real sign of a general rise. In Britain, for example, the rates fell during the two world wars and were particularly high during the economic depression of the 1930s. They were high again in the late 1950s. Currently, the rate is falling in older people but rising dramatically among young males. For example, between 1982 and 1992 the suicide rate for men aged eighteen to twenty-four rose by 85 per cent in England and Wales, far outstripping the rate among young women.

Suicide figures therefore suggest that depression isn't really on the increase except perhaps among young men. Unfortunately, suicide statistics are as flawed as any in their ability to tell us what is happening. First, there is no guarantee that the proportion of depressed people who commit suicide has stayed constant over the years. The current decline in suicide rates among older age groups in Britain, for example, is rightly or wrongly attributed to the success of modern antidepressants. Secondly, the borderline between accident and suicide is not always clear, and for many decades – and still in many countries – suicide was illegal or deemed immoral, pressurizing families and the issuers of death certificates into hiding ambivalent cases. These pressures may have changed over the years. There are certainly enough doubts over the data to stop anybody drawing a reliable conclusion about changes in rates of depression from changes in rates of suicide.

A Stigma in Decline

There is a stigma attached to mental illness. It is widespread among human cultures; enough for many people – and their families – to try to hide mental lapses such as depression. The usual ploy is to somatize the illness – to find physical symptoms on which to blame the mental problem. Nowadays in the West – and, increasingly, elsewhere – doctors are trained to look beyond physical symptoms in case there is a mental cause, another potential reason why the diagnosis of depression is on the increase.

The stigma of mental illness is so universal that it is probably part of our innate biology and as such can never be totally erased, or even completely masked. But just because something is ingrained or even innate does not mean it can't be strengthened or weakened by social or authoritative pressure. After all, the urge to breastfeed is innate – programmed into a woman's genes. But it can still be strengthened or weakened by social attitudes and situations. The same is undoubtedly true for the stigma attached to depression. Around Aristotle's time, for example, there was actually an element of social kudos in being depressed – at least for men of status – because it had already been noticed that 'melancholy' and creativity were often linked. By the 1300s, though, the Christian Church, irritated by the non-productivity of depressed monks, listed the illness as a punishable, cardinal sin.

In recent years, at least in the West, the traditional stigma attached to depression has begun to wane, thanks largely to an increase in public awareness. A survey in Britain in the late 1990s found that of more than 2,000 adults interviewed, 79 per cent considered depression to be a 'medical' not a 'mental' illness. Despite this, 51 per cent were afraid that their family doctor would think they were 'unbalanced or neurotic' if they went to the surgery complaining of depression. Evidently the stigma is still strong enough for half the population to be reluctant to admit to their condition. Nevertheless, people are less reluctant than in the past; the late-1990s UK survey found greater awareness of depression and more inclination to consult a doctor than an earlier survey in 1991. Such reduced reticence could produce an increase in the reported cases of depression without any real increase in the number of people actually suffering.

Interactions

A good example of the way the different factors – better or more cavalier diagnosis and reduced social stigma – can create the impression of a society with declining mental health is that of depression in young children. European and American governments alike have become concerned at an apparent increase in rates of clinical depression among children and adolescents. Other severe mental problems, such as manic depression or schizophrenia, are virtually unheard of in children and only rarely seen in adolescents, not developing until young adulthood. In the past, therefore, clinicians tended to overlook the possibility of related forms of such illness. There was also a particular stigma attached to childhood mental illnesses because children were simply too young for long-term environmental causes to be blamed. Consequently, young sufferers – unless they had just suffered bereavement or family upheaval – could only be considered as either genetically abnormal or potential victims of abuse.

As a result of these pressures, childhood mental illness was often dismissed in the past as temporary or not serious. The modern perspective on clinical depression, however, makes the illness among young children seem much more likely, leading to its more frequent diagnosis. Some child psychiatrists now routinely ask mildly depressed seven-year-olds about suicidal feelings, something unheard of around 1990. The result – of course – is that seven-year-olds are now more likely than in the past to voice thoughts of suicide, creating a real impression of deterioration in children's mental health. We simply cannot tell, though, whether this impression is real or merely due to a change in the readiness with which childhood depression has been sought and diagnosed.

Whether for children or adults, the absence of an objective test for clinical depression combined with changes in diagnosis and attitude prevent us from detecting real changes in incidence of the disease over the past few decades. For the moment, any claim that mental health is deteriorating has little solid foundation.

Depressing Genes

In all the hype, there is rarely mention of the genetics of depression – and when there is, genes are relegated to being just another risk factor. Everybody feels at risk – but the truth is, the risk is not equal for everybody. Moreover, scientifically, an understanding of the genetics of depression is absolutely crucial. Without it, no study of the causes of clinical depression is reliable.

Family Fortunes

Depression runs in families. If both parents have suffered from depression, a person has over a 50 per cent chance of also suffering – in other words, about four times the probability for the general population. If just one parent has suffered, a person has a 25 per cent chance of suffering – about twice the risk for the general population. When something clusters in families like this, it often means that genes and inheritance are involved – but there are other possible explanations. One is that an infectious organism, passed on by close contact, is the cause. Another is that the experience of being raised by depressed parents is a recipe for depression later in a person's life, irrespective of genetics.

The main evidence on the matter comes from studies of twins

and other non-twin siblings. Identical twins share all their genes; fraternal twins share only half – and non-twin siblings also share only half. So if a trait like depression is due to genes and not environment, any similarity between identical twins should be twice as great or twice as common as that between fraternal twins or between non-twin siblings. The data for depression show that if one of a pair of identical twins is depressed there is a 50 per cent chance that the other twin will become depressed as well. If one of a pair of fraternal twins becomes depressed, there is only a 25 per cent chance of the other twin becoming depressed. If one of a pair of non-twin siblings becomes depressed, there is again a 25 per cent chance of the other sibling becoming depressed. This is exactly the pattern expected if clinical depression owes everything to genes and nothing to shared environments (not even the mother's womb, as shared by twins). Adoption studies confirm this conclusion: an adopted person's chance of developing depression depends more on the depressive history of his or her absent genetic parents than on that of foster or adoptive parents. Genes and not examples seem to be the progenitors of depression.

Clinical depression is not the only mental illness associated with genetic make-up. Bipolar, otherwise known as manic, depression is an illness in which sufferers go through bouts very similar to those suffered during clinical depression. These bouts, though, alternate with euphoric periods of manic activity. Bipolar depression is even more strongly influenced by genetic make-up than clinical. If one of a pair of identical twins suffers from the illness, the chance of the other doing so is 75 per cent. Schizophrenia also has a strong genetic element.

Genetic Immunity?

Some people, therefore, carry genes that predispose them to becoming depressed; other people do not carry such genes. One critical but missing piece of information is whether the latter are actually immune to the disorder. We know, for example, that two-thirds of women cope perfectly well with exactly those stressful events that are claimed to trigger depression in the remaining third. Are these two-thirds genetically resistant to depression? We do not know – and we shan't know until a DNA test is available that allows people to be identified as genetically predisposed or not predisposed to depression, then seeing if any of the latter ever become depressed. If not, they may be immune.

This would be a vital piece of information. If some people are immune to depression, the nightmare vision of the whole of Western society functioning only on the back of antidepressants can be tempered. How much it can be tempered depends on what proportion of people are genetically immune. If it is the majority, then much of society will always survive untouched by such mental illness. If it is the minority, then the vision stands. Again, we have no DNA test that can tell us directly how many people are vulnerable and how many are not: but we can make a simple calculation.

Surveys suggest that between 10 and 17 per cent of people in Western societies will become clinically depressed at least once in their lives. As we have seen, such figures are unreliable and the range of uncertainty is likely to be much greater. For present purposes, though, let's assume a figure of 15 per cent, on the high side of the average. Studies of identical twins tell us that only half of people with the genes for depression (because their identical twin has suffered the illness) actually become depressed. It follows that the proportion of Westerners with such genes is double the proportion that actually suffer depression: in other words, 2 × 15 = 30 per cent. This would mean that 70 per cent do not have the genes and may even be immune. These are very important figures – but will always be imprecise until a DNA test is developed that can check them directly by randomly screening large numbers of people. For convenience, a split of 30:70 will be assumed in the sections that follow. The real split may be different – 20:80, perhaps, or 40:60. We can quibble over the numbers, but despite the imprecision, the principle remains: some people are predisposed to depression and others, probably the majority, are not.

The imprecision of these figures matters most when we try to understand differences between groups, such as between different cultures. For example, surveys in the Far East consistently give rates for depression that are half those in the West. We have already seen that we cannot trust such raw data. Stigmas are different in different countries and people may not feel equally free to admit their symptoms, even on anonymous questionnaires. But even if we take the figures at face value, we cannot interpret what they mean without knowing the genetics of the situation. Do only half as many people possess the genes for depression in the Far East? Or do equal numbers possess the genes in both cultures, but the Far Eastern lifestyle is only half as likely to push carriers into depression? We cannot answer these important questions without knowing more about the genetics of depression in both places.

Depressed Sex

The same problem applies when we compare groups within a culture: men and women, for example. Studies consistently show that women are much more likely to become depressed than men – but once again we cannot be certain. First, there are the usual problems with the raw data. Studies based on people seeking help – which usually put women at three times the risk of men – are particularly fragile. Maybe any stigma weighs less heavily on a woman than a man. As a result, maybe psychiatrists more freely diagnose depression in women, or maybe depressed women are more likely to seek help than depressed men. We can avoid these problems by using survey data rather than medical data. This makes some difference, but not much. Figures based on randomized questionnaires suggest women are at twice the risk of men – but now there is a new problem. Perhaps women are more likely to admit to depressive symptoms when filling in a questionnaire.

Surveys of bipolar depression – despite suggesting that the sexes are equally at risk – are just as unreliable and for all the same reasons. Even the statistics for suicide are of little help. Men are four times more likely than women to commit suicide. So are the surveys of depression wrong and in reality more men get depressed than women? Or are the surveys on depression correct – more women get depressed than men – but a depressed man is many times more likely to commit suicide than a depressed woman?

Secondly, there are the problems of interpretation. Even if we believe the surveys and accept that twice as many women get clinically depressed as men, what does this mean? Perhaps 40 per cent of women inherit the genes for depression compared with only 20 per cent of men. Such an inheritance pattern is possible, for example, if many of the genes for depression are inherited on the X-chromosome, of which women have two copies and men only one. Alternatively, perhaps 30 per cent of both sexes inherit the genes but female body chemistry makes depression more likely. Or perhaps a woman's lifestyle is simply more stressful than a man's and so more likely to trigger depression in those with the genes. Without knowing more about the basic genetics of depression, we simply cannot tell.

Anatomy of Depression

So perhaps around 30 per cent of people possess genes that predispose them to depression – but who are they? Can they be identified? DNA probes aren't yet available but maybe the genes for depression could be detected in other ways. After all, we know when a person has a Y-chromosome because he develops a penis and testes. Maybe the genes for depression also produce something obvious – such as a recognizable brain structure or chemistry, or even a particular personality.

At least two regions of the brain – the hippocampus and part of the cortex over the left cerebral hemisphere – have been claimed to be consistently smaller in depressed patients. In one study, for example, the hippocampus – essential for the formation of new memories – was on average 10 per cent smaller in women suffering from depression than in women who weren't. Moreover, the more episodes of depression a patient had suffered the smaller was her hippocampus. We don't know, though, whether each bout of depression caused hippocampal shrinkage or whether genetic depressives have a smaller hippocampus in the first place – and the smaller their hippocampus, the greater their predisposition.

It's relatively easy to imagine, rightly or wrongly, how a small hippocampus might predispose somebody to depression – smaller brains could mean fewer neurons and disturbed function. It's not quite so easy to imagine how depression might shrink the hippocampus, but there are two ways. First, the chemistry of depression – say a long-term production of stress hormones – could actually kill off brain cells. Alternatively, it could stop them from being replaced. Until recently, it was thought that all of a person's brain cells were produced early on in life, and that from adolescence onwards brain cells died at so many million per day, never to be replaced. Now it is known that new cells are formed throughout life. Depending on the ratio of cell death to cell birth, therefore, parts of the brain can shrink and expand during adulthood; shrinkage, for example, happens quite normally during pregnancy. Perhaps the chemistry of depression speeds up cell death or slows down cell birth, or both. Or perhaps genetic depressives have a naturally slower rate of brain-cell replacement throughout their life, even before their first bout of depression.

Chemistry of Depression

Maybe these studies show – though we can't yet be sure – that genes for depression have an influence on the size of some regions of the brain. Might the genes also influence the activity of various regions? Enter the PET (Positron Emission Tomography) scan into the field of depression research. This test involves a radioactive tracer to label a molecule – like oxygen or carbon – that freely enters the brain. The tracer is injected into a person and then followed by special cameras linked to a computer which reveal the parts of the brain which are most active and associated with particular mental activities.

The brain consists of an extraordinary number of nerve cells and connections – and to a large extent the development, inter-connection and activity of these cells are under tight genetic control. The nerve cells, called neurons, have long extensions from which they can receive inputs from up to 1,000 other neurons. Typically, though, each cell has only one long extension along which it transmits its own signal. There are an estimated million, million neurons in the brain. As we think, speak, move or feel emotions, neurons become active; electrical impulses flow through them. Damage the neurons, destroy some or all of their connections to other neurons or simply interfere with the electri-cal impulses in some way and the thoughts we have, the things we say and the way we feel can all change. The obvious example is the way that people's feelings, abilities and behaviour change in specific ways after localized damage due to a brain tumour or stroke. There is a language centre, for example, which if damaged disables a person's speech – and there are many memory centres. Damage one and a person loses some specific type of memory, such as the spatial information that allows him or her to find their way around.

Comparison of the brains of healthy and depressed subjects using PET scans has revealed a range of differences. In general, the research shows that a depressed state of mind equates with lowered brain activity, but there are other more specific differ-ences. For example, a part of the brain that in healthy patients is most active during moments of anxiety or sadness shows gener-ally greater activity in depressed patients. Another part, involved in specific cognitive tasks and emotional activities, has reduced activity in depressed people. On average, there is more right frontal brain activity in depressed people and more left frontal in non-depressed people. The pattern is consistent with the long-held view that the right side of the brain controls negative emotions, whereas the left side controls positive emotions. And so on – but as with the hippocampus study we don't really know whether these features of brain activity precede or develop as a result of depression.

Differences in brain activity between healthy people and depressives can be traced to differences in brain chemistry, such as in levels of hormones and neurotransmitters. Hormones are chemical signallers that course through the body in the blood and act over long distances. So hormones produced anywhere in the body can influence brain activity. Equally, the brain itself produces hormones that can influence distant parts of the body. Neurotransmitters are also chemical signallers but – usually – act over very short distances. Their main role in the brain is to allow neurons to communicate with each other across the gap between them, known as a synapse. Electric signals pass along the long extension to a neuron whenever the cell fires but they cannot jump the synapse to link up with the extensions of other neurons. Instead, the link is made chemically, the electric signal causing the release of a neurotransmitter, of which there are many different kinds. The neurotransmitter diffuses the very short distance across the synapse and causes the second neuron either to fire or to stop firing; some neurotransmitters excite neurons, some inhibit. Clearly, if any problem arises with a neurotransmitter – too much, too little, or some problem with the release, breakdown and re-absorption – the brain won't function properly.

The two neurotransmitters most often implicated in depression are noradrenaline and serotonin. Noradrenaline is associated with fight and flight responses and the regulation of arousal, pulse rate and blood pressure. Serotonin is associated with involuntary activities such as sleep, learning, appetite and libido – and is used by those parts of the brain also thought to be involved in depression. Serotonin, though, is found in places other than the brain – such as the gut and blood vessels. Practically, this is fortuitous. It is much easier to measure the concentration of chemicals in the blood than in the brain – yet the two levels often correlate. If somebody has a low level of serotonin in their blood, there is a good chance they also have a low level in their brain.

It might seem implausible that something as overwhelming and powerful as the way we think, function and feel could be changed by something so simple as the concentration of a chemical in the brain – but it shouldn't. Most people have experienced the buzz – and depressive after-effects – of at least one simple chemical, such as alcohol. Whether we like the idea or not our brains function on chemistry – and if the chemistry changes so do our thoughts, feelings and behaviour. Changes in serotonin level, in particular, can trigger changes in mood – though not dramatically. Lowering the level of serotonin in otherwise healthy volunteers (by asking them to drink a mixture that reduces serotonin synthesis) usually lowers their mood, though never to the level of a severe depression. Raising the serotonin levels of 'normal' individuals has no enhancing influence on positive feelings, such as extroversion and optimism, though it might reduce negative feelings, such as fear and anger. Studies of psychiatric patients have found various conditions to be associated with low levels of serotonin. People with a history of impulsive and violent behaviour are examples; so, too, are suicide victims. People with obsessive compulsive disorder (OCD) have reduced serotonin levels – and so, too, have people suffering from clinical depression.

Is it possible, then, that the genes for depression cause carriers to have a generally lower level of serotonin, and that this is the reason for their vulnerability? Maybe, but maybe not: the relationship between depression and serotonin level is not simple. Reducing the serotonin levels of depressed people still further does not make their depression worse. And antidepressants that almost immediately raise a depressed person's serotonin levels don't raise the sufferer's mood for three weeks or so, if then. In 30 per cent of cases such antidepressants have no effect whatsoever. It seems fairly clear that any link between serotonin levels and depression is somewhat different from the link between day-to-day changes in serotonin levels and mood. And as always, we cannot really decipher whether low serotonin levels are part of a depressed person's genetic predisposition, whether a drop in serotonin levels causes the illness, or whether the illness itself causes serotonin levels to drop.

Depressive Personality

There is still a great deal to discover about the links between brain structure, brain chemistry and brain activity and the way people think, feel and behave. The possibility is real, though, that those genetically prone to depression may have measurable features in their brains – or even blood – long before their first bout of depression and perhaps even at birth. So the final question is whether such innate features might also surface at an early age as personality traits.

Psychologists tease apart the human temperament into several independent spectra. The one that includes the depressive temperament is that of 'emotional sensitivity'. This spectrum has been studied in its own right, separate from specific considerations of clinical depression. In standard questionnaires, high scorers for emotional sensitivity are shy and have a deeply negative outlook on life, coupled with a prickly anxiety. Life is dark, the future's grim and every day's a chore. These people are potential depressives. In contrast, low scorers for emotional sensitivity are bold, calm and relaxed, with a positive attitude towards life and the future. Most people, of course, score somewhere between these two extremes.

One of the features of emotional sensitivity is how early it can be detected. A person's temperament in this respect may even be evident while still in the womb. A fast foetal heart rate leads to a whiny infancy, a shy childhood and an anxious adulthood; a slower foetal heart rate leads to a contented baby, a bold childhood and a confident adulthood. Many studies, following people through from childhood to adulthood, have confirmed that if a person is born shy or bold, the trait is likely to persist throughout life. Not surprisingly, therefore, a person's position on the spectrum of emotional sensitivity – not just their vulnerability to depression – is to a large extent determined by their genes. Studies of twins have shown that 50 to 60 per cent of the variation in emotional sensitivity between children is due to genes – and if only the extremes of sensitivity are examined, a massive 70 to 90 per cent of a person's temperament appears to be genetically determined.

Which Genes?

Knowing that there are genes for depression is one thing. Knowing precisely which genes those are and where they may be found in the human genome is quite another. And until the genes have been identified, it will not be possible to pinpoint who is vulnerable and who is not, what proportion of a population could ever become depressed, or even what really causes depression.

It is assumed that there are several genes involved in the predisposition to depression – one estimate, based on studies of emotional sensitivity, puts the figure at 11 to 14 – but none has yet been unequivocally identified. Two possible candidates, though, have so far emerged. American geneticists studying a gene named Ob, which produces a hormone involved in weight control, found the first. Some people carry shortened Ob genes that contain up to ten fewer DNA 'letters' than normal. The presence of two copies of a shortened Ob gene has been claimed to correlate with depression. The second candidate emerged from an international series of studies of a gene involved in the manufacture of a serotonin transporter, a chemical that helps serotonin move across synapses from one neuron to another. Once again, it is the length of the gene that seems to be important. People with a short version of the gene make less of the serotonin transporter than other people – and are more likely to be shy and anxious than bold and confident.

At the rate that progress is being made in understanding the human genome, it may not be long before the genes predisposing to depression are identified. Then the real study of the causes of depression will begin. In the meantime, the study of causation is fraught with problems.

Costs . . . and Benefits?

What causes clinical depression? In fact, there are two questions here – and neither has been answered convincingly. The first is why as many as 30 per cent of people in Western populations are genetically predisposed to depression. The second – discussed in the next section – is why only some of these genetic depressives go on to develop severe depression.

There are two schools of thought about genes for depression. One school, swayed by the debilitating nature of the illness, views the genes as faulty or defective. The other school, swayed by how common the genes are, views them as evolutionarily adaptive and hence functional. This second school, the school of evolutionary biology, claims that genes can only be considered faulty if they afflict less than 1 per cent of the population. Evolution would like to get rid of such faulty genes completely but can't because of the way natural selection works. So the genes for schizophrenia, which are possessed by about only 1 per cent of the population, can be considered faulty – but the genes for clinical depression are far too common and therefore cannot. However counter-intuitive it might seem, the alternative view has to be seriously entertained: genes for depression have a function that benefits either the depressive or his or her close genetic kin, such as parents, children or siblings.

If genes for depression have a function, why doesn't everybody possess them? Perhaps surprisingly, evolutionary theory does not always expect beneficial characteristics to be possessed by every individual in a population. Very often, evolution finds multiple solutions to a single problem. The obvious example is the problem of how to reproduce. Two main solutions have evolved: the male way and the female way. Both are successful and the human population contains two genetic types: men and women. This analogy may not seem to tell us anything about the evolution of depression, but it does. Usually, different genetic types evolve when each way of doing things has advantages and disadvantages compared to the other. Then evolution has a habit of fixing the proportions of the different types at the level that, reproductively, they are equally successful. In the case of men and women, the proportions have settled at more or less 50:50 (actually, 53 males for every 50 females). In the case of genetic depressives and non-depressives, the proportion seems to have settled – in Western populations, at least – at around 30:70.

The question, therefore, is what are the advantages and disadvantages of being a genetic depressive? It isn't difficult to see the disadvantages: long periods of reliance on other people, reduced productivity and a real risk of suicide. And as we saw when discussing cholesterol in the previous chapter, people who suffer from severe depression are also up to four times more likely to die from coronary heart disease than people who are not depressed. What possible advantages could offset such obvious risks?

The immediate temptation is to seek some paradoxical benefit from actually being depressed. One unconvincing suggestion, for example, has been that depressed people are showing behaviour that removes them as a threat from others higher up their social hierarchy. That way, the depressive has less chance of being attacked or killed. This may or may not be a reasonable function for troop-living primates but seems unlikely for humans. Another suggestion – made originally for post-natal depression in which mothers harbour negative or indifferent feelings towards their baby rather than themselves – is that a bout of depression tests the level of support available from a partner, extended family and wider society. Testing for support, though, is only functional if a person then acts on his or her conclusion once the test is over. Of course, if the support system passes the test, then the depressives can emerge reassured that their circumstances are supportive and continue their lives as before. But if the support system is found wanting, depressives may conclude that they need to change their life in some dramatic way. In the case of post-natal depression, a woman may be driven to abandon or even kill her baby. Infanticide by mothers presumably suffering from post-natal depression was rife in Western societies in the early 1900s and is still common in parts of the developing world. It was and is an extreme but biologically effective form of 'family planning', the woman sacrificing her current child to postpone parenthood until support and prospects seem better. Both post-natal depression and infanticide are much less common when new mothers have a high level of support. In the case of clinical depression, the role of infanticide is replaced by suicide. A frequent thought in the mind of suicidal depressives is that their family would be better off without them. If this were ever true – that close kin really would benefit reproductively from the depressive's death – then even suicide could be evolutionarily adaptive. Apart from being unpleasant, though, all such hypotheses are untested and probably untestable.

There is a much more appealing, though equally untested, possibility. The advantage of possessing depressive genes might lie not in the bouts of depression themselves but in the brain and personality of the depressive during their more buoyant phases. The risk of slipping into severe depression is the price a depressive pays for having their special type of brain and personality. So what might be special about the depressive mind? Many modern studies have confirmed statistically what Aristotle once suspected; the proportion of depressives – both clinical and bipolar – among writers and artists is significantly higher than average. Suppose this creativity springs from the depressive mind being exceptionally appraising, empathetic, introspective, romantic and imaginative as a result of its particular structure, activity and chemistry. This could well give depressives an advantage over their non-depressive counterparts for periods of their life; an advantage that in the long-term and in accordance with evolutionary theory just compensates for the disadvantageous periods of being depressed.

Direct evidence that the depressive brain might have special aptitudes is hard to come by, not least because most studies dwell on negative characteristics. One study, for example, reports that depressed people have had more nightmares during their life than healthy people. Depressives are also more likely to be hypochondriacs, though as always the data are subjective and based on people's self-appraisal and memories. Even such apparently negative indicators, though, might actually be signposting the heightened imagination and introspection of the depressive brain.

One of the most intriguing, albeit indirect, studies was of people falling in love. Investigators saw parallels between the obsessive behaviour in the early stages of a new relationship and the symptoms associated with OCD. OCD sufferers, like depressives, have reduced serotonin levels, so the researchers tested the serotonin levels of student volunteers who had just started a new relationship. Whereas the 'normal' students had the usual level of serotonin, the 'madly in love' volunteers had about 40 per cent less of the chemical. Then, to confirm that serotonin plummets solely during love's first flush, the researchers retested some of the original in-love students a year later: the students' serotonin levels had bounced back to normal, while a more subtle affection for their partner had replaced their original dizziness.

The startling warning posted by this research is that low serotonin levels can be advantageous. The ability to fall madly in love may not be possible without the chemical pathways needed for such drops in serotonin levels. So suppose the genes for depression endow a person with a generally lower level of serotonin, or make it easier for levels to fall. Might that mean that depressives fall in love more easily or have a more romantic nature? What other positive facets of depressive brain chemistry are waiting to be discovered?

Doctors, lawyers, writers and artists have an above-average chance of suffering from depression whereas people in managerial positions seem to have a below-average chance. This could mean that people with the genes for depression, because they are more empathetic, are attracted to the former jobs whereas people without the genes are attracted to the latter. Alternatively, the observed differences in the frequency of depression could be due to the unique stresses of each job. Maybe – as much of the hype suggests – it is life's tribulations from childhood onwards that determine who among genetic depressives actually becomes depressed.

Life's Traumas

Maybe evolutionary theory can explain why around 30 per cent of people in Western populations are genetically predisposed to depression – or maybe it can't. There is no certain way of testing the claims. Perhaps we have more chance of answering the second of the two questions about the cause of clinical depression: why only some – around 50 per cent – of those with a genetic pre-disposition go on to develop the illness. The expected answer is 'life'.

It sounds trite to blame depression on modern life, but that in effect is what the hype suggests about the role of our changing society and values. Two main types of cause have been mooted. One is the collection of traumatic 'life' events perhaps from early childhood onwards that might cause stress and indirectly damage brain chemistry. The other is the cocktail of prescriptive and recreational drugs and other chemicals that could directly damage brain chemistry. So are these really the factors that determine who among genetic depressives become depressed? Despite all we may read, little concrete progress has been made in producing an answer.

Stressful Events

First, we can consider the evidence about what are loosely termed 'life events' – traumatic experiences such as bereavement or a 'bad' childhood. These cannot be the only cause of depression because a third of all bouts have no such obvious precursor. But if they cause any, at what age are they most potent? The mean age for the onset of depression is currently under thirty years and seems to be declining. For people born since 1955, peak risk of depression is twenty-five years or under; for bipolar depression it is twenty-one. This implies that whatever triggers the illness happens commonly to young people – but how young? Perhaps, as with skin cancer, the situation is one in which events during childhood are critical. Maybe changes can be triggered in a child's brain that condemn him or her to depression later in life.

Certainly, a whole range of studies has linked adversity during childhood with an increased chance of some mental disorder later in life. For depression, the most consistent link is with lack of maternal care and the presence of family violence or sexual abuse. Marital discord between parents has also been claimed to be a factor. How much faith, though, can we have in such findings? Most such studies depend on asking patients to search their memories and dissect their childhood. And since the nature of depression is to be negative, it is inevitable that depressed people are more likely than healthy controls to seek and find negative elements in their distant childhood. And when we try to avoid this danger by limiting analysis to events – such as parental divorce, separation from parents or parental death – that are a matter of record rather than of memory or opinion, the evidence becomes much less impressive. Current data implicate divorce as a risk factor; also separation from parents, particularly a girl from her mother. Death of a parent, though, is not strongly implicated – in one study such parental death even seemed to reduce a child's risk of depression in later life. But even these data have problems. It is always possible that divorce or separation indicates that one or both parents was a depressive, diagnosed or not – with the child later suffering depression simply because it had inherited its parent's depressive genes. A study that followed the children of depressed parents into adulthood found three times the risk of depression, with peak age of the first bout coming between the ages of fifteen and twenty-five. This risk, though, is no greater than we might expect from inherited genes. And a study of female twins separated from their parents concluded that such experience had at most a minimal influence, calculated to be twenty times less important than genetic make-up.

So events during childhood may add little to the risk of clinical depression in later life. What about events during young adulthood, up to the mid-twenties? Unlike the situation for children, the clearest association is with bereavement, which is estimated to increase the probability of a severe depression in an adult by seven times. If depression strikes, it is usually during the first two years after the loss. The level of association is impressive, but not conclusive; there are still problems with such data. Although bereavement is a matter of record, its importance to the depressive is still a matter of the depressive's opinion. If we reduce this problem by breaking bereavement down into levels of relationship, the situation becomes far less clear. Death of a spouse, for example, may or may not be followed by depression. In one major American study some women suffered a depression that lasted many years after the death of a partner. Others appeared unscathed, even strengthened. The women who became depressed were more likely to say their marriage had been 'good' than those who did not – but women with a history of mental problems were actually less likely to become depressed.

Why isn't the pattern of depression in this and other studies as clear as we might expect if bereavement is such a powerful trigger? Perhaps the link between depression and bereavement works both ways, confusing cause and effect. In seeking to explain its state, the depressed mind might focus on the most traumatic recent event, of which bereavement of even a distant friend or relative is the most concrete. Perhaps a depressed person is seven times more likely than a healthy person to focus on a recent bereavement and claim to have been affected by it. In other words, a claimed association with bereavement could be the product rather than the cause of depression.

Cause and effect become even more difficult to unravel when we consider other types of life event – and for the same reason. A depressed person's own view of causation has no scientific validity; it is anecdote, not evidence. Memories are unreliable, and the depressed mind may distort events, exaggerating in order to make the depression seem more explicable. So it is not surprising that depressed people concentrate blame on life events with an element of humiliation, entrapment or death. Always there is an element of loss in their explanation, either a sudden loss – of a job, of a friend, of a precious idea – or a long-term loss, such as of freedom, health, a relationship or financial security. Such factors may well play a part in the triggering of depression, but the evidence is anecdotal and unreliable. Many people – including those with depressive genes – experience stressful life events without sinking into depression.

There is another, perhaps surprising, complication when trying to link life events with depression. Life events are not entirely random; some people are more prone to suffer trauma than others and in part the reason is genetic. Studies of twins suggest that about one-quarter of life events that twins both experience even when living apart are linked to their shared genes. How? Because events like criminal attacks, car injuries and industrial accidents that seem random are partly due to lifestyle, personality and drug and alcohol intake, which are not random and have a genetic element. Perhaps genes for depression predispose people to encounter – or even seek – stressful life events. For example, twins with depressive genes that attract them to drinking alcohol (as described later) may be more likely to have car accidents or provoke fights than twins without those genes. In which case any observed association between accidents and depression could be due to co-occurrence (that is, both caused by depressive genes) not causation (that is, accidents causing depression).

Chemical Assault

There is no question that depression can be precipitated chemically: the drug reserpine, for example, is known to be a depressant. More often than not, though, causal links are suspected rather than proven. For example, one of the syndromes shown by soldiers returning from the Gulf War involved an increased risk to depression, as well as slurred speech and other symptoms. For those involved, the war was undeniably a traumatic life event. But those most at risk to this particular syndrome were not the ones who had witnessed most horrors. They were either those who had worn a pet flea collar to ward off biting insects, or those who had been on night-time watch when insecticide mists were used to combat malaria and other insect-borne diseases. Another example of a chemical suspected of triggering depression is a contraceptive that delivers progesterone through six silicone rods implanted under the skin of the upper arm. Over 6 million women use the contraceptive worldwide, but it was withdrawn in Britain following a glut of bad publicity. Hundreds of women claimed to have suffered from irregular periods, hair loss and suicidal depression after receiving the implant, though a lawsuit against the distributors collapsed. Yet another suspected chemical link is with the long-term exposure of some farmers to pesticides in sheep dips. Suicidal depression, memory loss and sleep disturbance are just a few of the symptoms that have been attributed to the organophosphate insecticides used.

These are all possible examples of direct chemical triggers for depression, though none is yet proven. Other possible examples are more tortuous because cause and effect is yet again confused by genetics. Hypochondriacs, for example, are at an elevated risk to depression. But is that because they take more medications that might trigger depression chemically, or because – as considered earlier – people with genes for depression are also more likely to be hypochondriacs? Time and again, such confusion between co-occurrence and causation clouds the issue, as the following examples – drug abuse, aerosols and alcohol – illustrate.

Several studies of the use of the drug ecstasy have suggested a link with memory impairment and depression. PET scans show differences between the brains of ecstasy users and people who say they have never taken ecstasy. The researchers looked at synapses that specialize in releasing serotonin after subjects had been injected with a radioactive substance designed to 'light up' in the presence of healthy synapses. The control subjects had normal synapses, but the ecstasy users had deficiencies in all brain regions. Does this mean, though, that ecstasy damages serotonin transport, thereby triggering depression – or that people with the genes for depression have particular synapse characteristics and are more likely to be attracted to taking ecstasy?

Frequent use of air fresheners and aerosols in the home by pregnant women could be linked to post-natal depression. Air fresheners and aerosols – including deodorants, hair sprays and furniture polish – contain dozens of volatile organic compounds that can be toxic in high doses. In a large survey of thousands of pregnant British women, researchers found that those who used aerosols and air fresheners on most days suffered a 19 per cent increase in post-natal depression compared with those who used them less than once a week. But was this because aerosols and air fresheners cause post-natal depression or because a woman with genes for depression is more likely to obsess about cleaning, as in OCD, and so use more aerosols and air fresheners?

There is a long-standing link between alcoholism and depression. A large number of alcoholics show signs of depression and up to 25 per cent of alcohol and other drug abusers eventually kill themselves. Of teenage suicides in the United States and Finland, nearly half had been diagnosed as depressives and nearly a third were addicted to alcohol or other drugs. Biochemists report unusually low levels of certain messenger molecules or receptors in the brains of chronic alcoholics. So, do the genes for depression also chemically predispose a person to alcoholism? Or does depression drive a person to drink? Or does alcohol cause depression? North American Indians, for example, have a high level of depression linked to alcoholism. Most also have a genetic inability to metabolize alcohol, placing them at greater risk to alcoholism. Perhaps they also have a high incidence of genes for depression. Quite clearly, much more work needs to be done before we can unravel the true relationship between alcohol and depression.

These last few examples have illustrated a potential and confusing interplay between depression, genetic predisposition and 'unnatural' chemicals. There is also a potential interplay with 'natural' chemicals in the food we eat. In principle, a link between diet and depression is entirely reasonable, but again it is difficult to tease apart from genetic predisposition. For example, the amount of serotonin manufactured in a person's body depends in part on his or her diet – and a person's diet depends in part on genetic predisposition. Serotonin is manufactured from a very simple chemical, tryptophan, which is a normal part of the diet – in milk, for example. Work on rats has shown that foods rich in carbohydrate can increase the amount of serotonin made by the brain. So should we expect a low-carbohydrate diet to increase the risk of depression? There is no real evidence for this – though as we have seen there is some hint of a link between low blood cholesterol levels and suicidal depression. Once again, though, we cannot separate cause and effect; we cannot conclude that a person could avoid depression simply by increasing their blood cholesterol level. The link could be genetic rather than causal. Perhaps genes for depression also maintain a lower level of blood cholesterol.

Even if the chemicals we eat, drink and breathe do have the potential to cause depression, there is still a major difficulty in separating chemical causes from life events. After bereavement, people might turn to sleeping pills, alcohol, nicotine, and pain-killers – or even to eating too much of the wrong sort of food or too little of the right. How can we tell without properly controlled experiments whether bereavement triggers depression directly or whether bereavement triggers a new dietary and drug regime and it is the latter that chemically causes the depression? Similarly, might people who lose their job, are humiliated, lose their spouse or their child and yearn for social solace and support also seek chemical solace and support – and it is this that triggers their depression, not the loss? Maybe, also, people with the genes for depression are more likely than others to respond to bereavement or loss by turning to medications and alcohol. This will further exaggerate any impression that life events can trigger depression directly. No study has yet been able to tease apart the contributions of genes, life events and chemical intake – and until they do, we cannot be convinced that any of the mooted factors is a definite cause of depression.

Infectious Depression

There is one final complicating factor – disease. Long spells of illness, particularly if sleep is disturbed, have often been identified as a risk factor for depression. Depressed patients report more illness generally in their past than their healthy counterparts but, intriguingly, report fewer episodes of fever in the decade before their first bout of depression. Perhaps fever is unusually rare in potential depressives because they produce excessive amounts of stress hormones, which block the chemicals that trigger fever in response to infection. Alternatively, perhaps fevers protect against depression: raised levels of the appropriate chemicals boosting brain cells that use the chemicals dopamine or noradrenaline, some of which are thought to be sluggish in depressed people.

In theory, there could sometimes be a more direct effect of illness on brain chemistry than simply via the effects of stress. Many disease organisms actually change the body and brain chemistry of their host. Some do this directly via chemicals they release; some do so indirectly by simply tying up the host's immune system. Yet others do so because they chemically attack those organs in the host that secrete hormones that in turn influence the chemistry of the brain. Some may even attack brain chemistry directly. For example, a virus that causes bizarre behaviour in animals has been found in the blood of people with acute depression. Borna disease virus (BDV) infects many animals including horses, cows, cats and ostriches, and may cause apathy, hyperactivity or other behavioural disturbances. Studies in the United States, Europe and Japan have shown that about 30 per cent of people with psychiatric illnesses have traces of BDV's genetic material and proteins in their blood, compared with at most 5 per cent of healthy people. The virus itself has been isolated from the blood of a few people with acute depression. When the researchers injected these 'human' strains of BDV into rabbits, the animals became apathetic and stopped their usual grooming.

Whether BDV and other viruses can specifically attack brain chemistry to trigger depression or whether they simply add to a person's general stress is still unknown. But even if such viruses are a factor, they do not trigger mood disorders in everybody – a person must have some predisposition, such as a genetic predisposition, to succumb. The work on BDV does highlight, though, that for somebody with the genes for depression, a chance event like picking up a virus can trigger a series of events that eventually lead to a bout of depression. But we still cannot tell whether it is the disease organism, the stress of illness, or even the long-term effects of medication taken to attack the virus that actually causes the depression.

Treatments for Depression

It is clear that we still know very little about the causes of clinical depression. The only concrete knowledge is that two-thirds of people cope with traumatic and humiliating life events that if they happened to others would be linked directly or indirectly to a bout of depression. Perhaps these are the 70 per cent of people without the genes for depression. We have made little progress, though, towards understanding why only about half of those with the genes eventually succumb to the disease. Life events, medication, recreational drugs, diet and disease have all been implicated by research, but no study has yet been able to decide whether all of these are risk factors or whether one is far more important than the others. And the answer matters, because it not only influences our understanding of the causes of depression, it also influences our approach to its treatment.

There are, of course, still people around who believe that all a depressed person needs to do is take a cold shower and snap out of it. Now that we know, though, that depression is genetic and associated with changes in brain size, structure and chemistry, even the most die-hard member of the 'pull-yourself-together' brigade should be able to see that such a view is ill-informed, unsympathetic, and archaic.

Once a person has developed severe depression, there are four main forms of treatment. One is to do nothing and wait for the illness to go away. The others are various forms of active treatment, either some form of psychotherapy, drug treatment, or electromagnetic therapy. All have advantages and disadvantages; the main advantage of the treatments as opposed to doing nothing is that they help the patient to feel better and maybe function a little. The disadvantage is the danger of side effects.

Treatments and Side Effects

Doing nothing and waiting for the illness to go away is an option. Most bouts of depression are self-limiting; around 80 per cent of people make a complete recovery, the time needed averaging about six months. Doing nothing at least avoids the side effects of the more active treatments – but has the disadvantage that the person has to endure all the consequences of the illness itself and run the risk of suicide.

The least invasive form of treatment is psychotherapy, which in essence involves the patient talking on a regular basis with someone trained to talk and listen to depressives. The aim is to try to break the feedback element in a depressed person's thought cycle: depression leads to negative thoughts, which lead to stress, which leads to more depression. Improvement is usually expected to take about eight weeks. Exactly what the psychotherapist says in these sessions depends on the discipline he or she favours. There are no dangerous side effects from psychotherapy, other than perhaps the risk of dependence of patient on therapist. Partly to avoid this risk and partly to make psychotherapy more appealing to otherwise treatment-shy age groups – such as young men – a computerized Therapeutic Learning Program (TLP) has recently been developed. TLP is a self-help computer program in which patients work through a series of menus, first to identify sources of stress and relevant factors in their lifestyle, and then to work out what they could do to resolve the situation.

The biggest development in the treatment of depression in recent decades has been the emergence of whole ranges of anti-depressants. All are now thought to work directly or indirectly by increasing the concentration of neurotransmitters in the synapses between brain neurons. The first few generations of drugs, though, were accidents; developed as antidepressants after being observed to have such an effect while being used for the treatment of other illnesses. In that sense, they followed in the tradition of the oldest antidepressant, a folk remedy for anxiety, depression and sleep disorders, which consists of extracts of the herb St John's wort (*Hypericum perforatum*). This remedy is often pre-scribed in Germany and, because it has fewer side effects than modern antidepressants, is being increasingly prescribed else-where. The first modern drug for the treatment of a mental illness – largactil for schizophrenia – had its origin in the treatment of allergies in the 1930s. The so-called tricyclic drugs for the treat-ment of clinical depression were first tried out in the 1940s as a treatment for Parkinson's disease. At about the same time, MAOIs (monoamine oxidase inhibitors) were being used for the treatment of tuberculosis and were noticed to be anti-depressive. The most recent generation of antidepressants, though, are designer drugs, developed in accordance with modern theories of the chemical basis of depression. These are the SSRIs (Selective Serotonin Re-uptake Inhibitors), such as Prozac, with their roots in the 1960s and their flowering in the 1990s. They were devel-oped to block the normal uptake of serotonin in the synapses in the brain. Pharmacologists tried to synthesize a molecule so similar to serotonin that the neurons would take it up instead of the serotonin, thus leaving more serotonin outside the cell to transmit nerve impulses between neurons.

The main problems with antidepressant drugs are that they are slow to work and can have unpleasant side effects. MAOIs take six weeks before their effects begin to show; tricyclics and SSRIs take three weeks. In addition, early MAOIs led to liver damage and more recent ones may lead to dizziness, fainting, headaches and sleep problems. They also react adversely with certain foods such as mature cheese and pickled fish. Tricyclics can hinder urination and defecation, affect heartbeat, and lead to faintness, drowsiness and confusion. Even worse, for treating an illness so linked to suicide, overdosing is serious and can lead to death. SSRIs commonly induce nausea, insomnia, agitation and anxiety and may lead to sexual dysfunction, such as impotence in men.

A whole new set of side effects is produced when antidepressants are withdrawn. MAOIs can lead to headaches and nightmares. Tricyclics can produce anxiety, headaches, tremors, confusion, nausea and convulsions. SSRIs can produce delirium, nausea, fatigue and dizziness. For all of these reasons, the ending of an anti-depressive treatment is usually spread over a period of several weeks. Antidepressants are not thought to be addictive in the same sense as recreational drugs. However, it has recently been questioned whether the apparent relapse into depression that often occurs after medication stops is in reality a drug-withdrawal response rather than a return of the original depressive illness.

In part because of the side effects of antidepressants, there is an increasing reversion to one of the original treatments for major depression, Electric Convulsion Therapy (ECT). The idea may even go back 2,000 years to when an electric eel was placed on the head of a Roman emperor to cure his headaches – and was still in favour in Ethiopia in the sixteenth century. The first 'electroshock' of the modern era was administered in Rome in 1938 and quickly became the main medical treatment for the mentally ill, largely because there was little else available. The treatment, though, developed a bad public image. Side effects in the early years were extreme – severe memory loss, broken bones due to the muscle spasms, and, so it was claimed, occasional death. So when the first generation of antidepressant drugs came on the market in the 1960s, ECT fell out of favour.

In recent years, though, there has been a rebirth; ECT has begun to cast off its past image and return to fashion. In its new and improved form, supporters say it is a neat and clean medical procedure, with few apparent complications apart from some initial memory loss and confusion in the hours following the treatment. Modern ECT therapy uses anaesthetics to put the patient to sleep, muscle relaxants to stop muscle spasms and broken bones, and masks to make sure the patient gets enough oxygen. In all, the treatment takes about fifteen minutes from the time a patient lies down on the treatment table to when they start to wake up, but to be effective it needs to be carried out two to three times a week for several weeks.

It is still not known how ECT might work. Supporters suggest that the treatment in some way alters the volume of neurotransmitting signals that pass between nerve cells. Brain chemistry could be 'reset' following an ECT seizure, correcting the abnormal production of signals that control mood and behaviour. Sceptics claim that the treatment simply injures the brain so that patients can't remember what was stressing them. But with the American Psychiatric Association now saying that ECT can work for more than 80 per cent of severely depressed patients, it is increasingly being promoted as the most effective treatment available. Even so, in the United States the treatment is still primarily a last-ditch solution – but in Britain, it is increasingly the first treatment given to severely depressed, suicidal patients for whom a quick response is desirable.

Unlike ECT, TMS (Transcranial Magnetic Stimulation) isn't designed to cause seizures and convulsions. Rapid pulses of magnetism are delivered from a hand-held electromagnetic device the size of a baseball glove. Supporters claim that magnetic pulses are simply a clean and efficient way of 'injecting' small amounts of electricity into the brain. TMS does not cause a seizure because in contrast to electric currents, magnetic pulses pass straight through the skull. As a result, the brain is stimulated in a more focused manner, not with one large blast of energy but a series of gentler pulses. Nerves are turned on and off 800 times in twenty minutes, doing with repetition what is done by brute force with ECT; each magnetic pulse injects some 10,000 times less energy into the brain than does conventional ECT. There are still drawbacks, though. Existing electromagnets, although powerful, are only able to stimulate brain tissue to a depth of a few centimetres. This puts the brain's inner structures, many of which feature strongly in explanations of mood disorders, out of range for the time being. And the effects on the brain are still not as focused as some proponents would like.

There are a number of other subtreatments. Many people claim exercise to be beneficial, improvement occurring after about eight weeks. Manipulation of sleep patterns can also ameliorate the symptoms. Total and partial sleep deprivation improves around 60 per cent of depressed patients but the improvement disappears as soon as they sleep again. When depression is seasonal, as in seasonal affective disorder (SAD), the standard treatment is to expose patients to light from a sunlamp; the best time reported seems to be for an hour in the morning. It is not entirely clear why any of these treatments might work, but it is assumed that they in some way influence the release and synthesis of neurotransmitters.

But Does Treatment Really Work?

This is a difficult question. Part of the problem is the self-limiting nature of depression. As we have seen, even doing nothing produces an 80 per cent recovery. The proportion may even be higher than 80 per cent because an unknown number of people suffer depression but do not seek help, in part because they recover before doing so. This natural recovery system makes the appraisal of treatment difficult; the simple fact of recovery is not enough. Suppose a person was advised to talk to their pot plants every night as a form of therapy and after six months had recovered. Advisor and patient might well feel that the therapy had worked, whereas in fact, of course, the patient has simply recovered naturally and at the normal rate.

So the efficacy of a treatment needs to be judged less by whether patients recover or even simply feel better – compared with patients not being treated – than by the depth of their depression, their speed of recovery and their chances of suffering a further bout in the future. Comparison with people who receive no treatment is particularly important if the 'talking to a pot plant' error is to be avoided. The usual procedure is to compare the responses of a group receiving treatment with the responses of a control group receiving some form of placebo. Either a patient in a psychotherapy trial simply talks to a psychiatrist without following any set psychiatric protocol or a patient in a drug trial takes pills with no active ingredient. In ECT, control patients are given an anaesthetic but not the shock.

In study after study, 30 to 40 per cent of patients are reported to show a positive response to a placebo, whatever form it takes. What should we make of this? Many interpret it to mean that the simple thought of being treated is enough to trigger an improvement. But is 30 to 40 per cent really a positive response? How can we be sure this isn't just the pot-plant syndrome – people showing the level of recovery that occurs naturally? We can't measure this, of course. How can we quantify the responses of people who don't seek help and won't take part in a study? But if the six-month average and 80 per cent spontaneous recovery is true, we should expect roughly 50 per cent of people on placebos to have recovered after six months and nearly 80 per cent after a year. If anything, therefore, depending on timescale, a 30 to 40 per cent response to placebos could indicate an interference with recovery, rather than an enhancement. There are two main reasons why placebos could appear to interfere: the difficulty of measuring changes in level of depression, and the danger that placebos are not really true controls.

Measuring whether a patient's depression eases during treatment is not easy because, as in diagnosing depression in the first place, it depends on the patient's and clinician's subjective appraisal and/or scores on a questionnaire. How many self-reported symptoms have to disappear for it to be judged that the patient is responding to a treatment? Normally, the ad hoc level taken to reflect real improvement is half. Given such subjectivity, it is absolutely vital that placebo treatments really are placebos, with neither the clinician nor the patient knowing whether the patient is receiving proper treatment or the dummy. Otherwise there is a real danger that either the patient or the clinician or both will report, interpret and score according to the response they expect. And in the case of placebos the joint expectation is 'no response'. Part of the problem is often avoided by having a third party substitute for the clinician in scoring the symptoms, but there can be no substitute for the patient. Placebos are particularly difficult in tests to evaluate psychotherapy. The psychiatrist must know whether he is pursuing a particular discipline or merely chatting; hence there is a risk the patient will also realize. Placebos may even be impossible for drugs because the presence or absence of side effects may indicate to both patient and clinician whether a placebo or drug is being used.

So what can we deduce about the efficacy of different treatments? Probably very little – only two clear facts emerge. First, every form of treatment unambiguously fails in about 30 per cent of cases. Secondly, there is no real difference between treatments. Within psychotherapy, for example, the format doesn't matter; the response to group psychotherapy, for instance, is no different from the response to computerized TLP – even face-to-face communication is not essential. Similarly, whether the treatment is psychotherapy, exercise, antidepressants or ECT, studies consistently show that about 60 to 70 per cent of people appear to respond positively insofar as their mean score for symptoms decreases by a half. If we take a more stringent measure of response – say, no depressive symptoms at all for two months – then the rate falls to 30 to 40 per cent, with again no clear difference between treatments.

Do these figures mean that the treatments are all equally effective – or all equally ineffective? Even this is difficult to decide. One major problem is that all studies have their dropouts; up to a third of people fail to complete the course. If these dropouts were simply a random selection of participants, they would have no effect on the conclusion. On the other hand, if all dropped out because they were showing no response to the treatment then an apparent 60 to 70 per cent success rate is in reality only a 40 to 50 per cent success rate. Equally, if all dropped out because they were cured, the success rate is in reality a 70 to 80 per cent success rate. The success rate, therefore, lies between 40 and 80 per cent, and as such mirrors that for natural recovery.

Whatever the treatment, even if recovery is real it doesn't last long; many become depressed again. About a quarter of patients relapse within a year and about three-quarters will become depressed again within ten years. Treatments do not clearly differ in their prevention of further bouts.

Everybody involved in the treatment of depression, whatever their technique, feels that their own particular brand of therapy is effective to a degree. Similarly, the majority of patients (two-thirds) also feel that the treatment they are receiving, whatever it might be, is helping them at least a little. The overwhelming impression from the outside, however, is that there is no really effective and lasting treatment for clinical depression. Something is still missing.

Many believe the problem is simply the difficulty of matching treatment to patient – and much current research is aimed at remedying this. Maybe everybody would respond if they could be given the right treatment for them. If a clinician could predict who might and who might not respond to, say, Prozac it would be a big help, but only a little progress has so far been made. PET scans, for example, suggest that people who respond positively to antidepressants have a different blood flow in part of the brain from people who do not. Similarly, the number of rapid eye movements that depressed people make during the dream phase of sleep may be linked to whether they will respond to sleep-deprivation treatment.

Even if matching treatment to patient is the answer, there is still a long way to go before the treatment of individual depressives rises above the level of trial and error. For the moment, the impression is that the problem with treatment lies deeper than simple mismatching. It seems much more likely that the understanding of the illness itself is still flawed.

The Maybe Collection

Apart from the discovery that there is a genetic predisposition to depression, everything else is a collection of maybes. Maybe depression is on the increase, is more common in Western societies, and is more common in women – but maybe not. Maybe it is caused by life events, low serotonin levels, and prescriptive and recreational drugs – but maybe not. And maybe treatments such as psychotherapy, antidepressants and ECT help patients come out of depression or at least help them to feel better while suffering depression – but maybe not. Hard evidence is difficult, sometimes impossible, to obtain even in principle – and so far all we have are weak multiple effects that cannot be disentangled.

With so much being written about depression, we could be excused for thinking the illness is well understood. But this is far from true; the current pyramid of theory feels precariously like a house of cards. Two developments are crucial. First, we need an accurate and objective method of diagnosis – plus a measurement of severity that does not depend on the subjective judgement of patient and/or clinician. Secondly, we need a way of identifying people with a genetic predisposition without having to wait for them to become depressed. With these two developments, real progress could be made towards understanding causes, measuring incidence and appraising treatment. Maybe this would confirm all of the current theories; cement the house of cards into a solid edifice of understanding. But it would not be that surprising if instead it blew much of the old understanding away.

It is impossible not to draw parallels with the history of our understanding of stomach ulcers. For decades, these painful lesions were blamed – just like depression – on a stressful life and the irritating influence of chemicals, particularly alcohol and certain foods. Also just like depression, ulcers clustered in families. Treatment ranged from stress management to drugs and a change in lifestyle. The evidence on cause and effect was weak but 'made sense' and treatment seemed to work, after a fashion – though the ulcers usually returned. But when evidence is so tenuous and treatment so ineffective, it can mean that the basic understanding is flawed – and in the early 1990s, it was finally discovered that 95 per cent of stomach ulcers are caused by a bacterium, *Helicobacter pylori*. Nowadays, a course of antibiotics can clear the complaint within a fortnight – and it rarely returns. This is not to say that an organism – like BDV – also causes clinical depression. But maybe it – or something equally unexpected – just might be the decisive missing factor.

4

Mad Cows and Englishmen

In the spring of 1985, a cow on a farm in Kent, England, died a strange death and, over a year later, veterinary pathologists identified a new illness. Mad cow disease had arrived and before long had reached epidemic proportions. It was untreatable and always fatal; well nigh 200,000 cows had succumbed by the turn of the century and many times more had been slaughtered in an effort to limit the disease's spread.

The mad cow story has been so hyped it is widely known. The disease had originated in sheep and jumped to cattle, not naturally through hobnobbing on tranquil pastures but through forced carnivory. Cattle feed was to blame – protein food supplements produced by grinding up and pelleting the remains of disease-ridden sheep. But if the disease could jump along the food chain from sheep to cattle, might it not do so again and jump from cattle to humans via infected beef? In 1990, a poll showed that mad cow disease, which hadn't yet killed anybody, held more fear for 10 per cent of Britons than AIDS, which had already killed hundreds of thousands. Non-vegetarians everywhere – wherever British beef had been on the menu in the 1980s – held their breath, but not for long.

In 1996, the bad news came. Ten people had died of a human form of mad cow disease, presumably through eating infected beef. Nobody knew how easily the disease was transmitted or how long the incubation period might be. Uncertainty reigned. The gloomiest of soothsayers were suddenly warning that half the British population could already be infected. The whole of

Europe was also threatened, as was the rest of the meat-eating world. After all, who really knew whether they had eaten infected beef during the 1980s? As the twentieth century drew to its close, epidemiologists everywhere waited for the first signs of a human epidemic – and are waiting still.

This is the story, but how much is proven? Was mad cow a new disease? Did the feeding of infected sheep remains to cattle start the epidemic? Has the disease spread to humans from eating infective beef? Does half the human population of Britain – and others elsewhere – consist of the walking dead; infected people just waiting to stagger and die?

What Are 'Mad Cow' and Similar Diseases?

Mad cow and similar diseases form a group of illnesses that afflict a range of mammals, including humans. Technically they are known either as spongiform encephalopathies (SEs) or – more pronounceably – as prion diseases. The best known are scrapie in sheep and goats, chronic wasting disease (CWD) in captive deer, transmissible mink encephalopathy (TME) in ranched mink and – mad cow itself – bovine spongiform encephalopathy (BSE) in cattle. The current tally of prion diseases for humans is Kuru, Gerstmann-Straussler-Scheinker syndrome (GSS), fatal familial insomnia (FFI), and Creutzfeldt-Jakob disease (CJD). More may be waiting to be discovered. All are characterized by relatively long incubation periods, sometimes exceeding ten years. The brain becomes spongy in texture with fluid-filled cavities appearing in the tissue together with clumps of useless protein.

The first of the prion diseases to be widely studied was scrapie, which is endemic in the sheep flocks of Western Europe and has been known for 250 years. The brains of affected sheep show massive degeneration; the animals develop a staggering gait and other behavioural abnormalities before they die. BSE is a relative newcomer and in 1985 was the first example of a prion disease in cattle. Infected cows show no signs of disease for several years

– most are four or five years old when they first develop symptoms – but then the course of the disease is relatively rapid and always fatal. Affected individuals show characteristic and progressive neurological symptoms. They begin to lose coordination of their limbs and as their brains become irreparably damaged they become sullen, irritable and unpredictably aggressive.

The human prion diseases are all rare. Kuru, for example, is mainly localized in Papua New Guinea. In contrast, CJD is found worldwide but still affects only one person in about 1 to 2 million each year. The disease was first described in 1919, but attracted little attention until the late 1960s when experiments showed the disease could be transmitted to chimpanzees and was therefore infectious. Like the other SEs, CJD usually takes many years to incubate. When symptoms finally appear, usually in the sixth decade of life, the course is catastrophic. Illness begins with loss of memory and perhaps uncharacteristic behaviour, progressing to rapid mental deterioration, dementia, lack of coordination, and involuntary movements of the limbs. It is always fatal and most people die within a few months. The majority of cases are spontaneous, seeming to arise from nowhere. A small proportion clusters within families and a few other cases are the unfortunate result of medical mishap. Infection during brain surgery or corneal transplantation has caused some cases. So, too, has past treatment for dwarfism using extracts of pituitary hormone from corpses.

Although CJD is rare, its historical frequency may have been underestimated. A British study of over 1,000 preserved brains from people diagnosed as having died of dementia in the preceding twenty-five years found nineteen had in fact died of CJD. Only eleven, though, had been correctly diagnosed. Clinicians had labelled the remainder as having other diseases: six as having Alzheimer's. The symptoms recorded when the patients were dying were not typical of CJD: they had lived longer than usual after the onset of dementia and they had fewer signs of damaged coordination. However, even if half of all cases of CJD have been misdiagnosed in the past, even at two cases in every 1 to 2 million instead of one it is still a rare illness.

Spooky Prions and Bad Apples

In many ways the spookiest aspect of prion diseases when first studied was the ghostlike nature of the agents that caused them; none were visible and they were very difficult to destroy. They readily withstood the high temperatures, irradiation and doses of ultraviolet light of standard sterilization techniques – treatments that would kill any normal virus or bacterium. Only autoclaving at very high temperatures or high concentrations of sodium hypochlorite seemed to have any effect. Originally, the infectious element was called an 'unconventional virus'. But no virus – or bacterium – was ever found in infected brains. Finally, suspicion fell on the conspicuous clumps and plaques of useless protein that littered the diseased brains. Now it is claimed that rogue molecules of this protein are the infective agent itself, which has been dubbed a prion, shorthand for 'proteinaceous infectious particle'. Textbooks of infectious diseases needed a whole chapter to cover this new category of invader.

The protein that forms in the diseased brain is called Protease resistant Protein (PrP) because it is so resistant to attack by enzymes. It is now known, though, that PrP exists in two forms. There is a normal form, found in all cells in the brain, especially on the surfaces of neurons. Enzymes can break down this form. Its function isn't clear – suggestions range from neurotransmission, sleep regulation, and a role in immune activation to either promoting neuron survival or preventing neuron suicide. There is also a rogue form, the one found in spongy brains, that is subtly different and so resistant to enzymes that it accumulates, as if clogging the tissues.

The suggested course of infection is as follows. An animal swallows a prion and the invader is absorbed through the walls of the gut, eventually finding its way to the spinal cord. From there it slowly migrates up the nerve tracts to the brain where it begins to interact with the normal PrP. The normal proteins are corrupted, irreversibly, into rogues, a process that can actually be followed in a test tube. Corruption is a 'bad-apple' effect: the initial prion triggering a chain reaction in which neighbour corrupts neighbour until all the normal PrP has been converted. This eventually leads to the large clumps and plaques of protein characteristic of spongy brains.

So what is this subtle difference between normal PrP and the super-resistant rogue PrP that rots the brain? The favourite theory is that it is simply shape. The amino-acid chain in a normal protein folds into helical structures, though these exist in a variety of subtly different forms. In contrast, a rogue molecule is partly folded into a sheet. If this story is true, it is revolutionary: protein shape is enough to encode the biological information needed to cause an infection. 'Normal' viruses and bacteria need nucleic acid – either DNA or RNA – to do this job.

The 'infective-protein hypothesis' still awaits its final proof. Most researchers accept that rogue PrP is one component of the infectious agent but some continue to question whether it is the sole component. A few still expect to find an accompanying, albeit tiny, length of nucleic acid, which will be the real 'information' molecule. Packaged in a protective shield of rogue PrP protein, this nucleic acid would enter cells, shed its packaging and then go on to catalyse the conversion of normal PrP into rogue PrP.

The main reason such dissenters still expect to find nucleic acid is that there are a number of different strains of the best-studied prion disease, scrapie. These strains play an important part in the BSE story later on in this chapter. So far, around twenty forms of scrapie have been isolated; they can be identified by inoculating minced-up sheep brain into the brains of a 'panel' of five breeds of mice. The mice's responses to infection, incubation times, and the pattern of holes and tangled proteins in their brains give a consistent 'fingerprint' for each scrapie strain. Most importantly, the characteristics remain consistent as the disease is transmitted from one mouse to another. Somehow, the strains are imprinted with their own identity.

Supporters of the 'pure-protein' prion hypothesis claim that the different strains of scrapie are protein molecules of slightly different shapes. The molecules therefore attack different-shaped versions of normal PrP and hence produce different symptoms. Opponents say that the strains are easier to explain in terms of nucleic acid. The jury is still out on the true nature of the prion – but the final verdict will make little overall difference to the mad-cow odyssey.

Whereas scrapie, BSE and most of the other prion diseases are the result of infection, most – but not all – cases of CJD are not. CJD is primarily a genetic illness, the result of mutations in the gene that orchestrates the manufacture of normal PrP in the human brain. Twenty or more such mutations have been identified so far, and different mutations produce slightly different versions of the disease. The mutated genes encode for a less stable form of normal prion protein – a form much more likely to flip into the rogue form. People who produce such protein in effect produce their own 'bad apples', developing disease in the absence of any contact with outside infection. A few such cases run in families, the mutant gene being inherited; others arise by chance, spontaneous mutations corrupting the PrP gene within single cells. Such cells die, releasing rogue prion proteins and setting off a chain reaction throughout the brain. Nevertheless, once a person's brain has produced such a rogue protein spontaneously, molecules of the protein can act as prions and are capable of infecting others if they find a way of entering their brains.

Out of the Blue and Along
the Food Chain? Where Did BSE Come From –
and Where Is It Going?

BSE appeared as if from nowhere – and spread with lightning speed. The first case was reported in south-east England in 1985, followed by a few more in the same area in 1986. Then suddenly, it began cropping up all over Britain. Within five years, it had become an epidemic that had claimed over 20,000 British cows and was still claiming more at a rate of between 250 and 300 cases a week. Where had it come from? And, more worryingly, where was it going?

Jumping Prions: Is BSE Really Scrapie In Disguise?

The immediate question when BSE first appeared was whether it was a new disease or whether it was simply the old acquaintance scrapie in disguise. The scientists of the day deliberated and informed the British government of their decision. The feeding of scrapie-ridden sheep to cows was to blame; all over Britain, scrapie prions were surviving the rendering process and infecting the cows to which they were being fed. The suggestion seemed more than reasonable. After all, it was already known that scrapie could be made to infect rodents; the standard method of identifying the different strains of scrapie was to inoculate them into a panel of mice. So maybe scrapie could make the jump to cows via enforced carnivory just as well as it could to mice via inoculation.

Not only was the jumping-scrapie hypothesis scientifically plausible, it seemed irresistibly convincing to the layperson. Maybe there was something hideously attractive about the thought of cows eating dead, minced-up sheep. Whatever the reason, the jumping-scrapie hypothesis is still the explanation for the epidemic that most Britons remember and believe – and at the time it coloured not only the first round of public concern but also the first political initiatives. The solution to the BSE crisis was simple: ban food supplements containing sheep remains and within five or six years, the epidemic would wane and disappear. And, in 1988, this was the first step taken by the British government. Even at the time, however, it was clear that protein food supplements per se could only be part of the explanation for the sudden epidemic. After all, meat and bonemeal for cattle had been made from sheep carcasses since the 1920s with no hint of a problem. Another factor had to be involved – and that factor was thought to be a change in the process by which sheep were rendered down. Previously, rendering had been safe, using a solvent to extract tallow from carcasses. But in the late 1970s and early 1980s at just the right time to explain the outbreak of BSE, the process was changed. The solvent-extraction step was replaced by heat treatment, despite it being known that solvents reduce scrapie infectivity to mice whereas high temperatures (even up to 138 °C) do not.

So convincing was this hypothesis that nobody in Britain or Europe thought it worth doing the obvious test and deliberately trying to produce BSE by infecting cattle with scrapie. In a sense, though, there was little political reason for such an experiment. With the feed ban in place, there was nothing further to be done even if such a test did vindicate the jumping-scrapie hypothesis – and if it didn't, the ban had at least erred on the side of caution. Eventually, it was left to American researchers in Iowa to perform the experiment – and to give the scrapie hypothesis its first setback. In 1990, the research team injected eighteen calves with a pooled sample of scrapie-ridden brain tissue from sheep. Sure enough, the cows became sick after fourteen to eighteen months. They were lethargic, had difficulty moving their hind legs, and died within five months. But there were none of the behavioural traits associated with mad cow disease and post-mortems showed that the brain pathology was quite different. In effect, the cows had scrapie – not BSE. Even when the researchers repeated the experiment using particular strains of scrapie, they found nothing that resembled BSE.

Perhaps, though, scrapie had changed its characteristics – 'mutated' – as it jumped from species to species to become BSE. This had never been noticed with scrapie before, but maybe it was a feature of the prion that had become BSE. To test this, a wide variety of animals were infected with BSE experimentally – by inoculation – but when the BSE prion was inoculated back into mice, its characteristics were unchanged. The BSE agent seemed to maintain its identity on passage through all the species tested and at no stage resembled any known strain of scrapie.

Such evidence, of course, cannot prove that the British BSE prion is not a strain of scrapie in disguise. We could still argue, if we wished, that this one particular scrapie strain had escaped detection in sheep and goats. We could even argue that a known scrapie strain made the jump to cows but was triggered by the chemistry of its new host to 'mutate' into the BSE prion and so no longer be recognizable as scrapie. However, combined with later evidence that the change in rendering process in the 1980s had little influence on scrapie infectiveness, scientifically the jumping-scrapie hypothesis became very weak. Now it lives on only in the public memory. But if the BSE prion did not originate and spread as a jumping-scrapie agent, where did it come from? Whatever its origin, it was probably a one-off event. The way that the epidemic spread through space and time from south-east England suggests that all cases from 1985 onwards came from a single point source; an index case, probably in the 1970s or even in the 1960s.

An alternative to the jumping-scrapie hypothesis is that BSE is a rare but long-standing disease of cattle and that something special happened in Britain that led to the epidemic of the 1980s and 1990s. This is reasonable – especially if we compare BSE to CJD in humans. After all, most of the historical cases of CJD seem to be rare and spontaneous mutations, with no known cause. Yet once the rogue CJD protein is produced in a person's brain it is infectious both to chimpanzees (via inoculation) and other humans (via surgical instruments and inoculation). The same could be true for BSE. Perhaps, every so often, the spontaneous mutation of a PrP gene could produce a cow with BSE. Briefly, there is a risk of the infection spreading but before the 1980s that risk always ended with the death of the infected individual. BSE had never become an epidemic, any more than CJD had ever become an epidemic in humans.

Surely, though, if BSE had always been around – even if it were as rare as one in 2 million individuals like CJD – it would have been identified long before 1985 and would also have been found in countries other than Britain. Not necessarily: cattle in the early clinical stages of BSE may either have been misdiagnosed or simply not diagnosed. In some countries – France and the United States, for example – in which rabies was endemic, BSE could easily have been misdiagnosed. The early symptoms of rabies include weakness and general disability followed by paralysis in the hindquarters, causing very obvious staggering. And rabies is not the only disease to produce such symptoms – eating ragwort growing wild on pastures, for example, can give cattle 'staggers'. So even in countries without rabies – such as Britain – misdiagnosis was possible. As film of cattle with mad cow disease was increasingly shown on televisions around Britain, there were several reports from veterinary surgeons that they felt they had seen rare cases with such symptoms long before 1985.

Marginal evidence that BSE existed outside of Britain came from Wisconsin in the United States. In 1985 there was an outbreak of transmissible mink encephalopathy (TME) that seemed traceable to the feeding of mink with rendered 'downer' cows. Downer cows are those that suffer from paralysis and can no longer walk; about half have no conventional medical explanation and some could well be the result of spontaneous BSE. When the TME agent was experimentally inoculated back into cattle it caused BSE-like symptoms whereas scrapie – again – did not. But when BSE from the British epidemic was inoculated into American mink, although there were signs of neurological disease, the symptoms were quite unlike the clinical profile of TME. The changes in brain structure too – although similar to TME – were consistently different in detail. The same disease appeared when

British BSE was fed to mink. The most reasonable conclusion was that the outbreak of TME had been due to feeding mink with pelleted cattle remains contaminated with an American – or at least, not the British – strain of BSE. And after retracing the passage of downer cows from farms, to renderers and on to mink farms, it was calculated that around one out of every 27,500 downers might have been suffering from American BSE. This would give a figure of about one cow in every million carrying the American BSE prion in Wisconsin in the 1980s.

If BSE is a long-standing disease of cattle – or even if it really is an escaped but unrecognizable scrapie prion – the raw material for an epidemic could have been lurking for decades, if not centuries. The important question then becomes why BSE suddenly became an epidemic in Britain in the 1980s. Britain's farming practices were not unique and few things changed in the 1980s. Other countries had sheep with scrapie. Other countries fed meat and bonemeal to their cattle. Other countries may even have had sporadic cases of BSE in their cattle herds. Moreover, cattle – and meat and bonemeal – were exported and imported worldwide. Yet something happened in Britain in the 1980s that produced an epidemic when nothing similar had happened before or elsewhere.

Many suggestions have been made as to what might have happened to unleash BSE in Britain. Organophosphorus pesticides used to treat warble flies, changes in the role of knackers' yards, bacterial toxins, the transmission of prions by insects, the growing of rapeseed plants, protein poisoning in a drive to increase milk yield, black polythene plastic bags in rendering units, invasion of English pastures by new ragwort species, molybdenum toxicity, importation of the bones of exotic antelope or deer, shortage of copper in the bovine diet, auto-immune reactions, chemical spillage from a factory just five miles from the first case of BSE, and the bacterial transmission of prions have all been suggested. None, though, has any real evidence to support it and most cannot explain the way the epidemic seemed to originate from a point source and spread. There had to be a better explanation than any of these for the British BSE epidemic – and, in fact, there are probably three: the illegal use of bovine growth hormone; the feeding of cows with rendered-down cow remains; and the 'vertical' transmission of BSE prions from mother to calf.

BSE began in the south-east of Britain but within a year or so cases appeared suddenly throughout the country. A contaminated and highly infectious agent had been transported around the country, probably in the early 1980s. Hormones made from bovine pituitary extracts were widely, albeit illegally, used in Britain in the late 1970s and early 1980s. Their role was to boost milk yields and induce multiple ovulation in prize cows. Ironically, this practice coincided with a spate of children developing CJD following treatment for dwarfism with growth hormones from human pituitaries. Both practices were desperately unlucky to end up using contaminated pituitaries, given a disease frequency of only around one in a million. But as far as BSE was concerned, that bad luck could have been enough to amplify the naturally occurring cases, getting the numbers of infected cattle up to a level where an epidemic became possible.

Contaminated pituitary extracts could have set the scene for an epidemic, but were not enough on their own to produce a full-scale outbreak. Some additional amplification must also have occurred. This probably was the result of BSE prions being distributed in meat and bonemeal – but these were prions derived from other cattle, not from sheep. For a while – several years – after the initial outbreak, food supplements included rendered-down cattle, increasing numbers of which will have had BSE. Calculations based on the rate at which the epidemic progressed suggested that each cow with BSE that was slaughtered, rendered down, and then fed to other cattle infected between ten and a hundred others. In 1988, as soon as the danger was realized, the practice of feeding protein supplements containing the remains of cows (and, but probably irrelevantly, sheep) was banned.

This food ban was expected to bring a rapid end to the epidemic. With a four to five-year incubation period, the epidemic should have begun to wane in 1992 and should have ended completely by 2000. It quickly became apparent, though, that enforced cannibalism could not be the only means of BSE transmission from cow to cow. After correcting for under-reporting, the epidemic waned far slower than expected – over 30,000 cows born after the ban developed BSE. Initially, the blame was levelled at farmers illegally using up stockpiles of contaminated meat and bonemeal. This may well have been a factor – but could not be the full explanation. Cows that had never been fed contaminated food supplements were still developing BSE: the disease appeared in Europe in the offspring of cattle exported from Britain and on 'green' farms that never fed meat and bonemeal to their calves. The only reasonable explanation was that the disease was being transmitted 'vertically', from mother to calf.

In reality, this shouldn't have been too much of a surprise. Ewes had long been known to transmit scrapie to their lambs, either in the womb or as they gave birth. Scrapie could also be transmitted 'horizontally', from adult to adult – probably via flock members eating placentas or pasture that had been contaminated by placentas. Horizontal transmission is less likely in cows, which give birth in calving sheds under much more controlled conditions and so have little chance to eat each other's placentas. But vertical transmission was always a possibility – and proof was finally obtained in 1996. The study started in 1988, just as the government imposed its food ban. Epidemiologists followed the fate of more than 600 calves, half of them born to cows with BSE, the rest to cows of the same age from the same herd but which did not have the disease. Each cow was to be killed when it reached the age of seven – by which time it should have developed BSE if it was going to – or earlier if it developed signs of the disease. Eventually, the team had 546 brains to examine. Of the 273 born to BSE mothers, forty-two had BSE. Disconcertingly, though, of the 273 whose mothers did not have BSE, thirteen still developed the disease. The excess of twenty-nine calves with BSE gives a maternal transmission rate of about 10 per cent or more. Vertical transmission had been demonstrated, but an even more alarming possibility had been unearthed: cows with no sign of the disease may still be infectious.

The team that carried out this experiment blamed the thirteen infected calves born to mothers apparently free of BSE on exposure to contaminated food after the ban. But there was an alternative – that those mothers were silently carrying the BSE prion when the calves were born. This explanation would actually solve one of the long-standing enigmas about BSE: why the infected numbers of cows within any herd built up to around 10 to 15 per cent but no further. The original BSE epidemic had grown through increases in the number of herds infected, rather than increases in the number of cows within a herd. Now there was an explanation. Initially, contaminated meat and bonemeal infected cows – maybe all the cows in a herd – but they showed no symptoms of BSE, being slaughtered before any appeared. The reason for the delay in clinical signs past the age of slaughter could have been that the disease develops slowly if contracted via food. Alternatively, perhaps cows were already old before they'd received sufficient exposure: peak food supplement does not begin until the age of two years. Either way, these sub-clinically infected cows would give birth to calves with a 10 to 15 per cent chance of already being infected through vertical transmission. It was these cows – already infected at birth – that developed visible symptoms of the disease before the age of slaughter, hence becoming BSE statistics.

The possibility that sub-clinically infected cows could still be infectious could not be investigated until a test was available to identify the presence of BSE in cattle without symptoms. In 1998, with the development of such a test, a Swiss study showed that for every clinical case of BSE, 100 more cows could be silently carrying the prion; cows that would pass unnoticed through the slaughterhouses and enter the food chain. Tests on the apparently healthy members of herds in which BSE had occurred gave rates of about fourteen silent carriers per 3,000 cows; on the healthy members of herds in which no BSE had occurred, the rate was one per 3,000 cows. These are minimum rates, though, because the test can reveal the prion's presence only once it has multiplied in a cow's brain to sufficiently high levels.

By the end of the twentieth century, the number of reported cases of BSE in Britain was dropping by 40 per cent each year. The impression from official figures was that the epidemic was over. However, by then all cows in Britain were being slaughtered by thirty months of age as a matter of policy, before they had a chance to develop clinical symptoms. No tests equivalent to those in Switzerland were being carried out to see how many British cows were actually carrying BSE. Even now, the prions could still be rife in British herds and being transmitted silently from mother to calf. The real end of the epidemic may be much further away than reported cases of BSE imply. However, without meat and bonemeal to multiply the number of cases and with not much more than a 10 per cent vertical transmission rate from mother to calf, the disease should eventually but silently return to its pre-epidemic level.

Silent Cows: Epidemics Abroad

One of the mysteries surrounding the British BSE epidemic was 'Why Britain?' There was nothing obviously different about British farming practice, either legally or illegally. In fact, Britain was unusual among European countries in banning the use of bovine pituitary extracts, which may have caused the initial spread of the prion. There are two possible explanations. One is that it was simply bad luck; the other is that there was something about the British countryside that BSE found particularly suitable.

If BSE was a one-off event, whatever its origin, it was simply unlucky that it happened in Britain. If it is a long-standing disease of cattle at a level of about one in a million, again it was simply unlucky that contaminated pituitaries happened to be used for the manufacture of growth hormones and that meat and bone-meal were sufficiently contaminated to escalate the epidemic. It could have occurred anywhere – it just happened to be Britain.

The alternative – that there was something particularly conducive about the British countryside – is also possible. For example, sheep exported from Britain to New Zealand and Australia, neither of which has scrapie, nevertheless developed scrapie. However, no further sheep, not even the offspring of the infected imports, developed the disease. There were no obvious genetic or other reasons why the sheep should develop scrapie in Britain but not in the two southern hemisphere countries. Perhaps an unknown environmental factor is involved that makes Britain particularly prone to farmyard prions; and perhaps this factor was also proactive over BSE. This would help to explain why an epidemic occurred in Britain but not apparently elsewhere. Unless, of course, appearances are deceptive.

In the early 1990s, governments outside of Britain acted to ensure the BSE epidemic remained a British problem. In 1990, before vertical transmission of the disease had been demonstrated, the EEC banned the import of all British cattle born before the introduction of the British government's feed ban. Australia, Israel, the United States, Finland, Sweden and the Soviet Union also operated bans. Such moves were primarily to protect the cattle industry. However, Germany and France also banned all imports of beef, ostensibly to protect the public from infection, ignoring demands from the EEC to lift their boycotts.

The moves were only partially successful. By 1997, most countries in Europe had admitted to at least one case of BSE – almost all blamed cattle or feed imported from Britain. The official numbers of cases, though, were small, all dwarfed by the then British tally of 167,000. Switzerland had reported 230 cases, Ireland 214, Portugal fifty-eight, and France twenty-seven; even more extreme Germany had reported five, Netherlands two, Italy two, Denmark one, and Belgium and Spain, for example, none. The reality, though, must have been very different. As the chairman of the European Union's scientific advisory committee for BSE admitted, there had of course been cases of BSE in countries that claimed to be free of the disease. But instead of being reported, animals with strange, undiagnosed symptoms of the central nervous system were simply slaughtered and in many cases ended up on the beef shelves in supermarkets.

No one expected a Continent-wide BSE epidemic on the scale of that experienced in Britain. Forewarned was forearmed – but the official figures were too low to be credible. As many European scientists have pointed out, around 57,900 British cattle were exported to the Continent for breeding between 1985 and 1990 and lived on average for another two years. Based on the proportion of cattle of the same age in Britain that developed BSE, at least 1,688 of those exported cows must have become sick. They were not reported – so they must have entered the food chain. Even more cases must have arisen from infected meat and bone-meal imported from Britain, the feeding of which wasn't banned by the EU until 1994. A French parliamentary inquiry into BSE published in January 1997 claimed that up to 16,000 tonnes of British feed were imported every year by France alone before the French government banned it in 1989. Other EU countries carried on importing British feed right up until the EU ban five years later. Switzerland later reported many cases blamed on British meat and bonemeal; France and the Netherlands followed suit with a few cases. But there should have been many, many more all over Europe.

Until 1998, definitive evidence of BSE came only from the autopsy of clinical cases. If cases weren't reported, there was no autopsy and no confirmation of the disease. The development of tests for sub-clinical BSE brought the potential to change all that. Such tests allowed the Swiss, for example, to calculate how large their iceberg of sub-clinical cases might be. Eventually, the tests were also speedy enough for all detectable cases to be kept out of supermarkets. However, only the Swiss took immediate advantage of these tests. Even the British government claimed they had already done enough and that testing in slaughterhouses for silent infection was neither necessary nor cost-effective. Elsewhere in Europe, the intransigent attitude of many governments – 'we don't have BSE so we don't need to test to find out if we've got

BSE' – was only slowly eroded. In late 2000, though, the EU announced Continent-wide testing for BSE during 2001; soon we may really know the extent of Europe's silent epidemic. The first confirmed native-born cases of BSE in Germany and Spain in November 2000 suggest the disease may prove to be more widespread than has ever previously been admitted.

Further afield than Europe, the chances of a silent epidemic are much lower, but not zero. The United States, for example, imported 499 cows from Britain in the 1980s before Britain's BSE epidemic led to an import ban. It would have been really bad luck if any of these imports had silently carried the BSE prion, which then multiplied unnoticed through the use of meat and bonemeal in cattle feed. Around 400 of these cows died and may have been rendered down into feed before the United States banned the practice. The chances are that neither the British strain of BSE nor the putative American strain behind the TME outbreak in mink managed to spread sub-clinically through the American cattle herd. It probably isn't worth the expense of testing to find out. But then again . . .

Jumping Prions: Is New Variant CJD in Humans Really BSE in Disguise?

With the BSE epidemic at its height and speculation rife that contaminated beef could infect humans, the CJD Surveillance Unit in Edinburgh – established in 1990 – was on high alert for suspicious deaths. In 1994, fifty-five British people died of CJD, twice the number of deaths recorded in 1985. Was this the beginning of an epidemic? Or had greater alertness simply led to more frequent diagnosis? Then, over the next two years, ten victims of CJD showed such strange symptoms and such an unusual pattern of damage to their brains that alarm bells really began to ring.

Earlier strains of CJD normally had an incubation period of decades, so almost everyone affected was elderly: the average age of victims was around sixty-three. The ten new victims, though, were much younger. Aged between eighteen and forty-one their average age was only twenty-seven years. Unlike older sufferers, whose first symptoms tended to be forgetfulness and odd behaviour, the younger victims suffered from depression and anxiety. Also, the disease progressed unusually slowly, taking on average thirteen months rather than the usual six to kill. But the most worrying factor was the damage in the victims' brains and the larger very distinctive masses of protein, which were very reminiscent of BSE. The victims shared no common genetic predisposition to classical CJD, nor had any undergone brain surgery or treatment with contaminated growth hormone. The immediate conclusion was that, as feared, BSE had jumped from cattle to humans.

But had it? After all, scrapie didn't seem to have jumped from sheep to cows, despite the initial claims. Nor had scrapie jumped from sheep to people, so why should BSE? Mutton and lamb from scrapie-infected sheep had been on the human menu for at least 250 years. Yet none of the existing forms of CJD could be attributed to jumping scrapie. Cases of CJD and scrapie show absolutely no correlation; nor even do cases of CJD and the distribution of sheep. There are as many cases of CJD in Australia and New Zealand, where there is no scrapie, as there are in Britain and France, where scrapie is endemic. Icelanders probably eat more sheep meat (and sheep heads) than any other people. Thirty per cent of their sheep have scrapie, yet the incidence of CJD is no higher in Iceland than elsewhere. Likewise, Japan has a similar number of cases, yet Japanese people eat little sheep meat and what they do eat comes from New Zealand. So what reason, other than panic and pessimism, led people to expect BSE to jump to humans – and why should a new form of CJD, just ten years after the first outbreak of BSE in cattle, automatically be assumed to be a jumping prion?

Some of the evidence, though, said that BSE had done just that. The new variant of CJD (nvCJD) is far more similar to the prion that causes BSE in cattle than BSE is to any strain of scrapie. The two agents – for BSE and nvCJD – are sometimes indistinguishable. For example, material from the brains of three patients with nvCJD was injected into the brains of a panel of mice from four different strains. BSE has characteristic incubation times in these mouse strains that distinguish it from all strains of scrapie. Precisely the same incubation times were obtained with the nvCJD prion. The damage seen in the brains of the mice also matched the pattern that is typical of the BSE agent. They also seemed to share a biochemical signature. Further evidence of similarity was obtained when BSE was inoculated into rhesus monkeys. Like humans with nvCJD, the monkeys became nervous and anxious and on post-mortem their brain-damage pattern – numerous daisy-shaped deposits of protein surrounded by rings of tiny holes – was just like that in human brains damaged by nvCJD.

Not all of the evidence, though, shows a convincing similarity. When marmosets – another species of monkey – were infected with BSE, their brain damage was different from the damage to human brains caused by nvCJD. In-vitro studies have also raised some doubts. When rogue PrP is added to normal PrP in a test tube, the amount of normal PrP converted within two days is a measure of the susceptibility of the host to the prion. When rogue and normal proteins come from the same species or from species known to transmit prion diseases to each other easily, over 80 per cent of the normal PrP is converted, reflecting the host's high susceptibility. However, when the proteins come from species that do not transmit to each other, none of the normal PrP is converted. Both BSE and scrapie prions can trigger the conversion of human PrP in a test tube but neither does so efficiently – only about 10 per cent converting in both cases. Human PrP seems no more susceptible to BSE than it is to scrapie.

The evidence over whether BSE and nvCJD are caused by the same agent is therefore currently inconsistent. Even if they are the same, though, it doesn't automatically mean that nvCJD is BSE that jumps from cattle to humans. There is an equally plausible if macabre alternative: that BSE is nvCJD that jumped from humans to cattle and was then multiplied during the BSE epidemic. It so happens that before the outbreak of BSE, tests for various human pituitary hormones used antibodies obtained by inoculating human hormones into other animals. Just as the inoculation of human pituitary hormones spread CJD from human to human, it could also have spread CJD to these other animals. These were then destroyed at the end of their useful lifespan and their bodies – which could have contained nvCJD – were rendered down and turned into meat and bonemeal. The most impure human hormones were inoculated in the 1960s, a timing just right for silent amplification into the BSE epidemic of the late 1980s.

A human origin for BSE could be consistent with the timing of the epidemic. It could also explain why the BSE strain was never found naturally in other animals (except humans), and could explain why, when scrapie never spread to humans in 250 years, BSE seemed to spread immediately – because it was in fact a human disease. The logic is seductive – and there is only one reason to resist: nvCJD was not seen in humans until after the BSE epidemic had begun. However, as evidence this is very fragile; the moment an earlier example is suspected, the case against a human origin disappears – and by no means everybody is convinced that nvCJD did post-date BSE. A study of Britain's largest brain archive revealed unusual cases of human dementia dating from well before the BSE epidemic hit cattle. Most were misdiagnosed as Alzheimer's disease. Some of the victims were relatively young and in a third of the cases where CJD was misdiagnosed as Alzheimer's, the disease had followed the distinctive slow-developing pattern of nvCJD. In this form, the disease follows a similar course to conventional dementia, with gradual loss of memory and personality. Psychiatrists often find weird diseases that end in death. But usually no neurophysiology is ever done.

Maybe, therefore, rather than nvCJD being BSE in disguise, it could be the other way round: BSE is nvCJD in disguise.

Can People Catch New Variant CJD
By Eating Infected Beef?

If the BSE and nvCJD prions are similar, or even the same, how easily could they spread from cows to people? The most efficient way of transferring prions between species is to mince up the brain from an infected individual of one species and inoculate it directly into the brain of another. Scrapie and BSE can be made to infect most mammals – including primates – in this way; and so, too, can CJD. Inoculation of brain tissue is also the most efficient way of transferring prions between individuals of the same species. We know CJD has been spread from person to person in this way and it is possible that BSE was also partly spread by this means.

Another relatively efficient way of transferring prions from one individual to another within a species is by cannibalism. Sheep may contract scrapie from eating each other's placentas. Cattle may catch BSE by eating each other's remains in meat and bone-meal. Even for humans, cannibalism may be involved in the spread of Kuru. This prion disease is very rare worldwide, but its incidence was once high among tribes – such as the Fore – in the New Guinea highlands in which women used to handle and maybe even eat tissue from infected corpses during funeral rites. Kuru was much more common in women than in men from the same tribe.

The least efficient way of transferring prions between species is for one to eat the other. As far as we can tell, scrapie never infected humans – or cows – through the food chain. Even in experiments it has proved difficult to pass on the scrapie prions in food. Tests on pigs and chickens – which still receive meat and bonemeal in their rations – have produced no clear results so far. Nor have tests on marmosets. BSE, though, seems to be more virulent than scrapie. Mice have been infected with the British strain by being fed minced-up cow brains; so, too, have American mink. There have been other indications as well. In 1990, a Siamese cat was found to be suffering from a prion disease, which could have been BSE. It had trouble coordinating its limbs, but showed none of the aggressive nervousness characteristic of cows with BSE – 'happy but drunk' was the description given by one vet. Examination of tissue from the brain showed the sponge-like holes characteristic of prion diseases. Other pet cats were also found with prion disease; so, too, were several species of ante-lopes in zoos. If these were cases of BSE, contaminated food was the obvious potential culprit, though nothing was proved and some cases were enigmatic. For example, three red-necked ostriches from zoos in northern Germany were also identified with a prion disease. Ostriches are omnivorous and the birds had been given meat from German slaughterhouses. But neither BSE nor scrapie had been officially identified in German livestock at the time.

Clearly BSE can infect other species – if only mice and mink – in food. And if the BSE and nvCJD prions are similar – and particularly if BSE originated in humans – there has to be a danger that eating cows can infect humans. If so, which tissues from cows are most likely to be infectious? Minced brains would clearly be the most dangerous – and consumption was banned in Britain in 1989 – but what about the rest of the animal? Experi-ments inoculating a panel of mice with a whole range of cow tissues gave a clean bill of health to: muscle (the beef we eat), spleen, spinal cord, lymph tissue, and milk and udder tissue as well as uterus, sperm, and placenta. The results were reassuring for beef eaters everywhere – until the demonstration that cows could transmit BSE to their calves. Some tissues in the cow were clearly carrying enough BSE prions to infect calves, but not enough to infect mice. The hope is, of course, that if there are not enough prions in a tissue to infect mice via inoculation, there are not enough to infect humans via eating. For the moment, though, this remains a hope, not a fact.

So, theoretically, humans could become infected with BSE prions, and the assumption is that such infection would manifest itself as nvCJD. Has it really already happened? The first person with nvCJD showed symptoms early in 1994. By November

2000, eighty-four people had died of the disease in Britain. Apart from one seventy-four-year-old, all the confirmed victims were aged between ten and fifty-nine. In fact, as many as 90 per cent had been under forty years old, with people in their twenties at the greatest risk. Earlier in 2000, though, a baby born to a mother with nvCJD seemed also to have the symptoms. The possibility was raised that nvCJD, like BSE and scrapie, could be transmitted vertically from mother to baby, presumably while the baby was still in the womb.

Some of the early cases of nvCJD may have had special histories: dairy farmers, for example, and a bodybuilder who may have used contaminated steroids. But by and large, there was no clear risk factor among the victims' medical histories. For example, there were no cases recorded as a consequence of medical or surgical intervention. In the absence of any evidence to the contrary – but without any positive evidence either – the assumption was made that the 1990s outbreak of nvCJD was the result of the victims eating contaminated beef products in the early 1980s.

If this is true, half the British population – and many others in Europe and elsewhere – could be incubating nvCJD. Clearly, from the recent evidence on silent infection, people have been eating meat from infected cows ever since the early 1980s. And until tests are carried out on all cattle in all slaughterhouses – tests that are fast and sensitive enough to keep contaminated meat off supermarket shelves – people will continue to eat infected beef. How many contract nvCJD as a result will depend on whether and how easily BSE really does infect humans through the digestive tract. In most parts of Europe, even clinically infected cows may still be finding their way to the dinner table. At least British government legislation catches these animals, even if it doesn't divert the silent carriers. The biggest question at present is how many cows are silently infected and whether such animals contain enough prions to infect humans. If they do, the next question is whether even the new generation of tests for BSE are sensitive enough. Could cows that pass these tests nevertheless still be carrying enough prions to be infective?

New Variant CJD: Optimism and Pessimism

Nobody knows how many people are really incubating nvCJD – and the current state of knowledge does not allow us even to guess. The most optimistic scenario is that the nvCJD prion is not the same as the BSE prion and cannot be caught by eating beef. In which case, the current spate of nvCJD cases is simply due to greater interest in prion diseases and greater alertness. Cases will not increase in number, yearly, and will trundle along in Britain at the rate of ten to twenty each year. The most pessimistic scenario is that the two prions are one and the same, that nvCJD can be contracted by eating beef, and that although government measures around Europe will keep clinically infected meat off our dinner tables, silent carriers will prove just as infective. In which case tens of thousands of people, if not half the meat-eating population, could currently be incubating prions in their brain. For the moment, science is unable to separate optimism and pessimism.

If we are looking for reassurance, the signs are that nvCJD did not increase significantly between 1995 and 1999. Three deaths occurred in Britain in 1995, ten in 1996, ten in 1997, seventeen in 1998, and fourteen in 1999. Although twenty-four had died by November 2000, the numbers are still nowhere near even the fifty or so new cases per year of classical CJD. In contrast, within six years of the first reported case of BSE the epidemic was nearly at its peak and the rate at which new cases were being reported had increased several thousand-fold.

On the other hand, if we are looking for reasons to panic, the time taken for human prion diseases to develop varies enormously. For Kuru, it is between five and thirty years, for classical CJD between eighteen months and twenty-five years, and for nvCJD – who knows? The first cases of nvCJD could simply be the very fastest developing ones with the massive bulk of the epidemic just waiting in the wings. Exposure of the British public to the most heavily infected meat probably peaked in 1989 and 1990, when the first moves to protect humans began to bite.

Guidelines were given to manufacturers of medicines that use bovine materials and, more radically, a total ban on the human consumption of certain cattle offal came into force, including a ban on their inclusion in baby foods. The consumption of meat from silent carriers, though, still continues. It is impossible to say whether the peak in any subsequent epidemic has passed or is still to come.

The need is for a test, one that could screen large numbers of people cheaply and easily. So far, none exists – though the drive is under way. The ideal is a blood test. PrP is carried on the surface of white blood cells, and so it may be possible to detect nvCJD prions in a blood sample. It is hoped that such a test will be ready by 2002 – but for the moment we have to make do with a test that is still less than user-friendly. Used initially on autopsied brains, the test was later found to detect rogue PrP in the tonsils, as well as the spleen and lymph nodes. In classical CJD, rogue PrP can't be detected in these tissues. The hope is that tonsils will test positive within months of infection of nvCJD, giving early warning of an incubating disease and, if enough people are tested, a future epidemic.

By 2000, the national CJD Surveillance Unit in Edinburgh was using an antibody test to detect prions in stored tonsils and plans were afoot for anonymous testing of 2,000 fresh tonsils removed during surgery. This is a start, but even if tens – or even hundreds – of thousands of people already have nvCJD, this would not necessarily be detected by such a small sample. For example, the first results were published in April 2000 – and the good news was that no signs of nvCJD were found in 3,000 tonsil and appendix specimens removed in operations since the 1980s. Even as the results were announced, though, the research team warned that the results were not an 'all clear'. The tonsil and appendix survey will eventually examine tissue from around 18,000 samples, but even this is not enough. A much larger screening programme is needed – and for this a blood test is the only feasible method.

This will not, though, be the end of the matter. A blood test draws attention to the fact that the nvCJD prion can be detected in blood. Might it not also be spread via blood transfusions? The panel of mice say that it can't – but they also said it was unlikely cows could pass prions on to their calves. In 1999, Canada and the United States banned blood donations from people who had spent more than six months in Britain in the 1980s. This produced a 3 per cent drop in donations, but allowed the Canadian health ministry to claim it had virtually eliminated the chances of nvCJD infection via the blood supply. Soon, there is bound to be huge pressure for the routine screening of donated blood, at least in Britain, so that it can be destroyed as a precaution if positive. But it could be too late; perhaps nvCJD has already been silently multiplied, just as BSE must have been in the early 1980s.

Probing the Genes

Among all the gloom, there is a modicum of good news, at least for part of the population. Genetic predisposition is probably a factor in the contraction of nvCJD, just as it is for depression and many other illnesses. In even the most pessimistic scenario, some people should be immune. Everything depends on the form of the normal PrP in a person's brain. All mammals – except for certain laboratory strains – have a gene for encoding this protein. Different forms of the gene, though, manufacture different forms of PrP, which in turn differ in their vulnerability to prion attack. In fact, mice lacking the PrP gene altogether are totally immune to injections of scrapie-infected brain tissue – because they have no PrP to be attacked.

In some mammals, both resistant and vulnerable forms of the PrP gene have been identified. In the late 1980s, for example, a simple blood test was developed that could discriminate between sheep that were genetically resistant or susceptible to scrapie. In reality, the difference is not actually one of resistance and susceptibility. All sheep will incubate the scrapie prion once they have it. The real difference is that in one type of sheep the scrapie incubation period is short, so the animal dies; in the other type, incubation takes longer than the animal's lifespan. The animal carries the disease, but silently, throughout life – and dies before it is killed.

Similar differences seem to be present in people. Of those who developed CJD after receiving grafts of brain tissue, 86 per cent had a particular form of the PrP gene. The same genetic trait is present in most people who develop CJD spontaneously. Intriguingly, a large proportion of people who developed CJD after receiving hormone injections – contaminated material entering the body through the bloodstream rather than the brain – also had an abnormality in the prion gene, but it was a different abnormality. Evidently, the two genetic traits predispose people to CJD in two different ways. Now, it appears that there is also a genetic link between the victims of nvCJD, but here the good news ends – because the same genetic trait is shown by 40 per cent of the human population. This may be heartening for the other 60 per cent, if only they knew who they were. But it also means that if an epidemic has been silently developing over the past two decades or so, its potential victims amount to nearly half the population.

The Long Wait: In the Shadow of Fragile Science

The British BSE epidemic and the nvCJD scare in its wake are the perfect examples of the modern vulnerability to fragile science. From the very beginning, vital political decisions had to be made on the back of very shaky scientific hypotheses just waiting to be disproved. The jumping-scrapie hypothesis, for example, coloured government actions around the world for the best part of a decade. It influenced decisions that ranged from how to control the British BSE epidemic to how necessary it was to protect humans. The scrapie hypothesis was almost certainly wrong, but it no longer really matters. Prion science has moved on to fragile pastures new.

What really matters now is whether BSE can afflict humans. Is nvCJD the human manifestation of BSE and – a different question – can it be caught by eating infected beef? If the answers are yes and yes, however, there is a sudden desperate need for prion science to start answering life-and-death questions with robust evidence, not fragile hypotheses. The conclusion that nvCJD can be caught from eating beef will crumble the moment anybody finds an alternative explanation for the cases so far. One alternative – suggested in the report of the British Government's official inquiry into the BSE epidemic, published in October 2000 – was that the disease might have been contracted through immunization. This could explain why most nvCJD victims are young adults. Some vaccines are made using culture media derived from cattle tissue. Although drugs companies were asked to get this material from BSE-free herds as early as 1989, they were not legally required to do so until March 2001. At least one company continued to use bovine serum from British herds – to make an oral polio vaccine – after receiving the early guidelines, and such vacccines could still have been in use until November 1993. The British government's decision late in 2000 to compensate the families of nvCJD victims might seem tantamount to an admission of responsibility for transmission of the disease. If so, it is a political rather than a scientific admission. Scientifically, it is still far from proven that infection with nvCJD is the result of exposure to cattle products, either beef or vaccine.

Wider questions also need more solid answers. If nvCJD prions can be detected in blood, can the disease be passed on in blood transfusions? Political angst may only just have reached the starting gate. Large-scale screening for a looming epidemic would create many problems, just as HIV screening did in the 1980s. Should medical authorities attempt to trace those who test positive and tell them the terrible news, or keep it to themselves? What would any of us gain other than the knowledge that we face an extremely unpleasant death? Currently, there is no treatment for nvCJD. There are no drugs to prevent prions from clogging the brain. And knowing that we carry the nvCJD prion is unlikely to help us prepare for our demise – especially if we don't know whether we have two years left to live or twenty. Knowing our 'prion status' might conceivably allow us to reduce the risk of infecting others – but how? There is no evidence – so far – that nvCJD can be transmitted during, say, sex, or via the minor mingling of blood that extra vigilance might prevent.

When the BSE epidemic began, the scientific advice to the British government was that, although there was a theoretical risk of the disease being transmitted to humans, the danger was acceptably small. The science on which that advice was based was fragile. When the first cases of nvCJD were reported a decade later, British beef became non gratis everywhere from British schools to the supermarkets of the world. Again, the science on which the panic reaction was based was fragile. Every single scenario from the worst to the best can scoop up some bit of evidence from which to claim support. But when push comes to shove, uncertainty is the only truth – because science simply does not have robust answers to the important questions.

5

Global Warming: Cheap Holidays, Future Apocalypse or Just Hot Air?

Good news! The world is getting warmer and the tropics are heading for the poles. The bleak northern – and southern – winters are soon to be banished to the textbooks of climatic history. By the end of the twenty-first century, heating and clothing bills will be slashed and there will be no need to fly away for that holiday in the sun. Tropical paradise will be just a bus stop past suburbia. Even better news; we can speed up the arrival of these halcyon days. Not only is global warming gathering momentum but we humans are helping it along. Gases we place in the atmosphere are acting like a blanket and keeping in the earth's heat. Industry contributes most – but we can all play our part in a concerted drive for a warmer planet. Let's burn more fuel and cool more fridges – and with the aid of confidential hotlines, let's point the finger at neighbours caught in clandestine acts of energy conservation.

Of course, this isn't the hype surrounding global warming, though it could have been. However, most scientists, politicians and conservationists in the developed world foresee a darker future. Depressing visions range from simple resignation – humans abuse Mother Nature yet again – to outright apocalypse. Polar ice will melt, sea levels will rise and the world's weather will become wild. Imagine the Statue of Liberty up to her neck in water and tropical storms playing dominoes with the New York skyline. Among the mayhem, tropical diseases will spread north and south and threaten us all. And while temperate countries cope with the downside of tropical life, their equatorial counterparts will have to cope with an ever-expanding desert: a band of sand from France to South Africa and encircling the globe. Three hundred and fifty million people could be at risk of famine.

So which is the scientific reality of global warming? Will our great-great-grandchildren live in a climate very different from ours? If so, will the change be a balmy source of pleasure to them, or will it make their lives hell? Will they thank us for making it happen or blame us for causing it in the first place?

How Bad Could It Be?

Potentially catastrophic: at least according to the scientists at the forefront of global-warming research, those who advise governments and feed the media. The United Nations Intergovernmental Panel on Climate Change (IPCC) consists of a consortium of over 2,000 climate scientists and its mandate is to assess and advise on research and policy options. It should be objective and in possession of all relevant information, and it has decided that global warming is a reality. On the basis of a 0.6 deg. rise in temperature over the past century, they predict that global temperature will have risen by between 1.5 and 4.5 deg. by the year 2100. The range of uncertainty is only 3 deg., yet the difference between the extremes is critical. The lower end of the IPCC prediction – a rise of 1.5 deg. – actually raises few problems. It would, in fact, simply return us to the balmy days our antecedents enjoyed from the tenth to fourteenth centuries – very pleasant for northerners and no global catastrophe. The upper end, though – a rise of nearly five degrees – would be a very different matter.

The last such major change was at the end of the last ice age. During the ice age tiny changes in the earth's orbit around the sun reduced the amount of solar radiation the earth received and the world cooled enough for vast ice sheets to creep south from the Arctic. It took a rise of 5 deg. to reverse the process; a rise that was enough to catapult high-latitude climates from being permanently buried under glaciers – or at least cloaked by tundra

– to the relative comfort enjoyed today. Imagine yet a further rise of the same magnitude. What could that do to our climate, the planet's biology, and human lives? The answer depends on latitude – but the calculations have been done and the predictions made. Wherever on earth we live, the results could be extreme.

In the tropics, global warming could mean drought and famine. As land and air temperatures increase, the atmosphere holds more moisture. An increase of 4 deg. in air temperature means that around 30 per cent more moisture evaporates from the ground, drying it out. Admittedly, there will also be an increase in evaporation from the oceans and hence an increase in global rainfall. But the increase in rainfall will be only 12 per cent – not enough to make good the 30 per cent loss from the land, so the land will dry out. Four degrees, then, would be enough to bring frequent droughts to middle latitudes and arid climates would extend about 35° north and south of the equator. One biological result would be a decrease in crop production – and risk of famine. A 10 per cent decline is being forecast in the production of wheat, maize, soya beans and rice in developing countries. Estimates for the numbers of people who in 2060 could be affected by famine range from 50 million to as many as 350 million. This is on top of a baseline population at risk of hunger, which will already have been swelled by population growth to some 640 million.

Paradoxically, significant tropical warming will not only bring drought but also increased hazards from severe storms and hurricanes, which feed on the energy unleashed as water vapour from warm oceans condenses into rain. A multitude of factors influence hurricane development, including temperature, wind and moisture; it is impossible to predict in detail how climate change will affect the pattern. But since tropical storms form only at temperatures above 26 °C, it seems likely that warmer seas will fuel more of them and that they will be more intense – and so wander further into temperate latitudes, affecting more people.

Sea levels will also rise – partly because water expands as it warms and partly because more of the polar ice will melt. In the early panic over global warming, sea-level rises of as much as 8 metres were being bandied around – enough for the heads of island states to talk about their nations being drowned and for Americans to plan to move the Statue of Liberty to higher ground. But the first major international conference on the greenhouse effect put the figure at a more conservative 80 centimetres and, in 1990, the IPCC acknowledged that the true figure would fall somewhere between 31 and 110 centimetres with a 'best guess' of 65 centimetres – inconvenient for most rather than cataclysmic, though some coastal farmers in China, for example, may suffer badly.

The flooding of oceans with freshwater from melting ice, though, may do more than simply raise the sea level and tinker with coastlines. It could have surprisingly dire consequences for Western Europe: global warming could paradoxically plunge most of the area into a virtual ice age lasting hundreds of years, bringing with it wider effects on climate that might be felt all round the world. Even more alarmingly, such unpleasant refrigeration could take place quickly, over a decade or so, and in our lifetimes. There would be little time for preparation – or migration. Lifestyles would have to adapt virtually overnight. How could such an unlikely catastrophe happen?

Europeans tend to forget that their clement climate is due entirely to second-hand warmth from the tropics being brought to them for free by the Atlantic Ocean Gulf Stream. By rights, northern Europe should experience the same chilly climate as the northern United States and Canada, since they are all at more or less the same latitude. But warm surface waters originating in the tropics are drawn northwards and cajoled into heating Europe. They release heat into the northern atmosphere at a rate of a trillion kilowatts (10^{15} W), an amount equivalent to one hundred times the world's energy consumption. This free energy warms the air over Europe by about 5 deg. and has operated reliably since the end of the last ice age. But why should this stability be threatened by global warming?

The only reason the Gulf Stream travels far enough north to

heat Europe is that the North Atlantic is home to two powerful oceanic 'pumps'. One of these is to the east of Greenland and the other is in the southern Labrador Sea; it is these – as they drag the Gulf Stream from the surface down into the deep ocean – that give the warm surface water from the tropics an extra tug to the north. The Gulf Stream and its pumps are part of a global circulation system that has been dubbed the 'conveyor belt'. If the belt stops, waters that will cool rather than heat Europe will replace the Gulf Stream – and temperatures in Europe will drop by about 10 deg. We know this could happen – because it has happened in the past. Studies of the layers of year upon year of snow that have piled up on the Greenland Ice Sheet and year upon year of sediment that have built up at the bottom of the Atlantic all show that rapid and severe climatic jumps occurred every 1,000 years or so during the last ice age. The last of these jumps took place as the earth finally emerged from the big freeze. Gradual climatic warming following a slight shift in the earth's orbit was causing the huge continental ice sheets to melt and disintegrate. But then, within a decade, ice-age conditions suddenly returned. All it takes, it seems, is for the North Atlantic to be diluted beyond a certain point – by extra rain or snowfall or by melting polar ice – and the conveyor belt switches off. Perhaps global warming could dilute the North Atlantic by just enough for this to happen again.

Tropical famine, hurricanes all over the industrial world, coast-lines changed and an ice age in Europe: no wonder the world panicked when the first descriptions of global warming hit the headlines in 1988. Science has posted its warnings; the consequences of ignoring them would seem dire. All over the developed world, governments have acknowledged how bad things could become and many are doing what they can to make sure the cataclysm never happens.

But does anybody really need to do anything? How good is the science behind the gloom? There are three main questions that we need science to answer: 1) is the earth really warming up?; 2) how much of the change, if any, is our fault?; and 3) can we – or should we – change the future, anyway?

Is the Earth Really Warming Up?

Maybe – certainly temperatures measured at ground level have risen over the past century. The increase was 0.3 deg. between 1890 and 1970 and a further rise of 0.3 deg. from 1970 to 2000 with the 1990s being the warmest decade on record. This might not sound much – but it's enough for the IPCC to extrapolate into the future and see catastrophe. However, this surface warming does not seem to be penetrating the atmosphere. Over wide areas of the planet, at heights above the earth even as low as 3 kilometres, the temperature has been at least stable if not falling.

Does such a contradiction between surface and atmosphere matter? We can answer this later. But as the surface warming alone has been enough to generate panic, we can begin by asking . . .

But Is It Really Our Fault?

Without an industrial human in sight, the earth's temperature has managed to rise and fall many times since the planet's formation. Three million years ago, for example, although the tropics were much the same temperature as now, the higher latitudes were 10 deg. warmer. The last time the earth cooled – as we have already seen – was during the last ice age, around 20,000 years ago. However, since the last ice age ended about 10,000 years ago, the climate has been remarkably stable. The earth warmed by about 1 deg. over about 400 years in medieval times between 950 and 1350 – the most balmy weather regime in recorded history. And in a mini ice age in the seventeenth century it fell by a similar amount.

What caused the changes on these occasions is unclear – but it wasn't human activity. Evidently, therefore, the earth has been quite capable of warming up and cooling down without any help from us. So what has happened that suddenly we deserve the blame for the modest changes witnessed at the planet's surface over the past century?

Gaia's Greenhouse: The Earth Before *Homo*

In theory, industrial humans could change the climate – because our chemical outpourings could interfere with the way the earth has always balanced its thermal books in our absence. The main buzzwords in the global-warming saga are 'the greenhouse effect' and 'greenhouse gases'. They sound innocent enough – in fact, they sound positively atmospheric – but in the past decade science has managed to imbue them with sinister overtones.

Gas and Glass

The earth is warm for the same reason that the inside of a greenhouse is warm. First, sunlight strikes the ground and warms its surface. Then, over time the ground radiates the energy back into space. As this outgoing infrared radiation passes through the atmosphere, a variety of molecules – the so-called greenhouse gases – hold it for a while. They grow warm and send infrared radiation back toward the earth, adding to the warming at the surface. The molecules have the same effect as the panes of glass in a greenhouse – and the warming process has been called the greenhouse effect.

The atmosphere may act like greenhouse glass but it is more complicated than the average windowpane, even the double-glazed variety. It is multilayered and each layer is physically different. The sun's light has to penetrate each in turn on its way in and infrared radiation has to escape each in turn on its way back out. From the outside in, these layers of the atmosphere are called the thermosphere, mesosphere, stratosphere and troposphere. The top layers of each are termed the '-pause': stratopause, tropopause, etc.

The troposphere is the layer of the atmosphere closest to the earth and the one in which we live. Its thickness varies from about 16 kilometres over the equator to about 10 kilometres over the North and South poles. The layer cools with altitude – reaching a low of about −55 °C at the tropopause – but it is a turbulent place. Air masses rise and fall as they heat at the surface and cool aloft, thereby generating nearly all of the earth's weather. Any gas in the troposphere that hinders the flow of heat back into space is a greenhouse gas. There are lots of them and they are virtually all totally natural, in the sense that they were present in the atmosphere long before humans were even a twinkle in evolution's eye. Carbon dioxide is the most common greenhouse gas but nitrous oxide is another, and so, too, are water vapour and ozone. By and large, the more greenhouse gas there is in the troposphere, the warmer the earth will be – though nothing in the global-warming catalogue is quite that simple.

There is nothing sinister about the greenhouse effect or greenhouse gases. They are part of the earth's natural state – and have been a vital element in evolution ever since our planet first acquired a gaseous atmosphere. We owe greenhouse gases a great deal. Without them, the surface of the globe would be a freezing −20 °C, so the oceans would have frozen and no life would have developed. As it is – thanks to greenhouse gases – the temperature of the earth at its surface averages a pleasant 13 °C. Inevitably, though, the temperature varies from time to time – just like the inside of a greenhouse. As every gardener knows, the alignment of his building, the strength of the sun and the thickness and quality of the greenhouse glass all influence the temperature inside. So, too, does how clean he keeps the windows.

Naturally Dirty Windows

A greenhouse with dirty windows is cooler than one with clean – because less sunlight gets through. Similarly, the dirtier the earth's atmosphere, the less sunlight also gets through – so less heat is radiated back from the earth's surface and the less the greenhouse gases can do their warming work. So, did the earth always have clean windows until humans came along and sullied the sky with industrial outpourings? Not really. The natural world has plenty of ways of blackening the atmosphere without human help.

Chemicals that darken the atmosphere, even those that have nothing to do with humans, are known – in the jargon of global climatology – as 'aerosols' and they help to cool the atmosphere in two ways: directly and indirectly. The direct effect is the dirty-window effect – to scatter sunlight and reduce the amount that hits the earth. There is also an indirect effect, which we can consider later when we wrestle with the tricky role of clouds in the earth's thermal budget.

There are two main size-classes of aerosols – molecules, which are too small to see, and particles, which aren't. At a molecular level, the most effective aerosols are sulphates – and the natural world has plenty of ways of sending sulphates into the sky. One of the main sources is marine phytoplankton – microscopic plants (algae) that live in the surface layers of the world's oceans. These tiny organisms produce a compound called dimethyl sulphonio-propionate (DMSP), which helps them to survive in their salty ocean habitat without dehydrating. When the plankton die, their DMSP is released into the water, where it breaks down to form dimethyl sulphide (DMS), which ultimately forms a natural sulphate aerosol in the air.

Major volcanic eruptions emit both molecules and particles, belching both many kilometres into the air, right to the very top of the troposphere or even into the stratosphere. Here they can stay for a year or more and spread right round the globe. Natural fires also dim the sun, producing a variety of aerosols including fine black soot. Dust, too, darkens the sky, particularly when storms cross major deserts. The largest sources are the Sahara, the deserts of Arabia, India and Australia and the Gobi of central Asia. And dust travels. Each year, trade winds carry more than

200 million tonnes of dust from the Sahara westward across the North Atlantic.

Cools v Warms: Earth's Thermal Tug-of-War

In the complex world of global warming, analogies help. The greenhouse analogy makes the basic principle of global warming seem simple. A further analogy can even make the dynamic contest between greenhouse gases and aerosols seem manageable: the two are like the teams (the Warms and the Cools) in an eternal tug-of-war contest, each trying to pull the earth's average temperature in opposite directions. When the Warms are winning we have global warming; when the Cools are winning we have global cooling. Nothing about the earth's climate, though, is really simple – or even manageable. As the first hint of complexity in the global-warming story, it turns out that some members of the opposing tug-of-war teams keep swapping sides; some even pull for both sides at once.

Ozone is one such duplicitous player. In the stratosphere it pulls for the Cools. Lower down – in the troposphere – it is a greenhouse gas and pulls for the Warms. Ozone is a form of oxygen but its molecules contain three oxygen atoms instead of the usual two. Before humans existed, ozone was produced from ordinary oxygen in two ways – via high-energy sunlight in the stratosphere and via electric discharges such as lightning in the troposphere. The stratospheric ozone pulls for the Cools by blotting out around 95 to 99 per cent of the UV rays heading for the earth. The much less concentrated ozone molecules in the troposphere form a greenhouse gas and so pull for the Warms. The net effect of ozone on the thermal tug-of-war, therefore, depends in part on whether its stratospheric persona pulls for the Cools more strongly than its tropospheric persona pulls for the Warms.

If the ambivalent role of ozone is a team manager's headache, the multifaceted role of water vapour is a nightmare. Water molecules in the atmosphere are a standard greenhouse gas and pull for the Warms. But excessive water vapour has a nasty habit of forming clouds – and clouds in the atmosphere are a bit like having a greenhouse roofed with mirrors pointing outwards. They reflect the sun's heat back into space before it has a chance to warm the ground beneath. Result: cloud formation can suddenly convert water vapour from a Warm to a Cool. But even this fickleness isn't the end of the matter. Which side is actually favoured by water vapour depends not only on whether it forms clouds but also on how patchy or continuous those clouds may be, how thick they are, how high in the atmosphere, whether they are over the tropics or the poles, and whether they are most common by day or by night. The details don't really matter here – it is enough to know that water vapour's contribution to the contest is patchy, erratic and probably totally unpredictable.

Left to its own devices, water vapour is simply fickle. Aerosols, however, do not entirely leave water vapour to its own devices. We have already seen that aerosols are Cools because they dirty the earth's atmospheric windows. But they also have additional – indirect – effects due to their influence on water vapour and clouds. Again there are many ways the two can interact. As just one example, when aerosols enter a cloud, each particle usually becomes the nucleus for the formation of a new water droplet. But there is only so much moisture in the air, so the more aerosols there are the smaller the droplets. Small droplets lead to whiter clouds and whiter clouds reflect more sunlight away from the earth. Result: support for the Cools. Almost every influence of aerosols on water vapour works in this way. Not only do aerosols pull vigorously for the Cools themselves, they are forever doing their best to encourage water vapour to pull with them rather than against them. Desert dust is a particularly loyal Cool. When in the air, it is a true aerosol and both dirties the atmosphere and encourages water vapour to defect to its side. Its contribution to the tug-of-war, though, doesn't end there. When dust falls from the skies, it often recruits a substitute for itself. Hundreds of millions of tonnes of iron-rich desert dust fall into the oceans each year, effectively peppering the oceans with fertilizer. Iron encourages the growth of DMSP-producing plankton. These eventually produce atmospheric sulphates, which give a further tug for the Cools on the demised dust's behalf.

After the capriciousness of some greenhouse gases – particularly ozone and water vapour – aerosols seem refreshingly loyal in the thermal tug-of-war. But sadly, as always, if we look hard enough we can find at least one minor indiscretion. In the dazzling white of the snow-covered Arctic, where most of the sunlight is normally reflected, the dominant direct effect of aerosols is not to reflect incoming sunlight, but to trap longer-wavelength radiation on its way back out. In spring, aerosols over the poles can actually give a tiny, surreptitious tug for the Warms.

Then Along Came Humans

Gaia's thermal tug-of-war has been acted out virtually ever since the earth was formed. And since plankton filled the seas and a cloak of vegetation spread over the earth's surface, the main players – greenhouse gases and aerosols – have changed very little. Since humankind arrived on the planet over a million years ago, though, we have increasingly irritated the earth's atmosphere. And when we invented pollution, we gained the power to give one of the two sides a decided advantage. But which side?

Humans and Global Cooling

Human action bolstering the Cools is taking place all round the world. The culprits range from the operators of power stations in Europe to goatherds on the fringes of the Sahara – from farmers burning Amazon rainforests to the iron and steel manufacturers of the Far East. All are filling the atmosphere with extra aerosols, over and above those that the earth produces naturally.

The biggest single source of human-generated aerosols is probably the sulphates produced by power stations, though aerosols from the burning of biomass may almost be as great. The burning of crop stubble, forests and other vegetation has increased steadily for more than a century. Today crop burning is commonest in the savannah regions of tropical South America and Africa, while forest burning has been most intense in the Amazon basin. Deliberate burning of grassland and forest spews around 50 million tonnes of organic aerosol into the air each year, compared with only 4 million tonnes from natural fires. Dust storms are another natural phenomenon that human activity has amplified. From ancient China, through the American Midwest half a century ago, to Africa today, over-farming and drought have led to the widespread destruction of soils, allowing their residues to be propelled into the air on winds. Estimates of the quantity of minerals rising from the world's windblown deserts range from 1 to 4 billion tonnes a year, though the fraction stemming from human activity is uncertain.

Even indirectly, humans might be increasing the air's aerosol content. Fertilizers, which provide nutrients for phytoplankton when washed into the ocean, could be increasing the amount of DMS in the seas and hence the concentration of aerosol sulphates in the atmosphere.

Humans and Global Warming

Humankind's major contribution to the Warms is thought to be our influence on the amount of carbon dioxide in the atmosphere. This gas is, after all, the most important greenhouse gas – the heavyweight anchorman for the Warms. With unanimity unusual in global-warming science, everybody agrees that the amount of carbon dioxide in the atmosphere is steadily increasing. We pour the gas into the atmosphere daily, mainly via the burning of the fossil fuels – coal, oil, and natural gas. Not only this but also, as discussed in the next chapter, our destruction of forests that naturally absorb the gas further acts to increase carbon-dioxide levels. The gas's concentration in the atmosphere today is nearly 360 parts per million (ppm), compared with 315 ppm in 1958 when modern measurements started and 270 ppm in pre-industrial times (as measured by air bubbles trapped in the Greenland

Ice Sheet). The IPCC prediction is that our production of carbon dioxide will double by the year 2100.

We are also releasing other greenhouse gases into the atmosphere, including methane, nitrous oxide and, indirectly, ozone. The human influence on ozone gives a double-tug for the Warms. As we have seen, ozone has always pulled for both sides in the ancient tug-of-war. Tropospheric ozone pulls for the Warms while stratospheric ozone pulls for the Cools. But because human pollution is relentlessly acting to strengthen the tropospheric ozone and weaken the stratospheric, we are siding with the Warms twice over. What is happening is this. In the troposphere, chemical reactions between sunlight and a variety of our pollutants are adding to the ozone naturally produced in this layer. Spawned by pollution, ozone itself becomes a pollutant – a component of smog. It also becomes a greenhouse gas. But while industrial and domestic processes are increasing the amount of ozone in the troposphere, the same processes are decreasing the amount in the stratosphere. As early as the mid-1970s, concern was expressed that chemical compounds called chlorofluorocarbons (CFCs) were breaking down the ozone layer in the upper atmosphere. The result was first manifest as a dramatic 'hole' in the ozone layer above Halley's Bay in the Antarctic. At that time, CFCs were widely used as propellants in aerosol spray cans. After CFCs are released, they slowly rise into the atmosphere. When they reach the stratosphere, the sun's ultraviolet radiation breaks them apart. Some of the molecular fragments that result react with ozone, thereby reducing the amount of ozone.

And the Outcome Is . . . Confusion

In principle, therefore, human activity has the potential to favour both sides of the tug-of-war. So which side are we really favouring: the Warms or the Cools? Or are we being even-handed and favouring both, and therefore neither? There is no answer. Some scientists think we are polluting our way to climatic catastrophe, others that our influence on the earth's climate is minuscule. Some think we are favouring the Warms, others the Cools. Why can't people at the forefront of climatologic research agree over such an important and fundamental issue? But then again, how could they?

The only hope for predicting the human contribution to future changes in climate is first to understand our contribution to past changes. And the only way to do that is to produce an all-embracing, all-seeing computer model that – had it existed in 1900 – could have predicted the ups and downs of the climate over the past century. No other method exists beyond hanging up seaweed or reading tea leaves.

To work, such a computer model needs to be able to recreate mathematically the thermal tug-of-war. It needs to be able to quantify the pull of the various Warms and Cools and express the extent to which they pull in opposite directions. It needs to allow for vagaries such as those that pull for both sides, those that keep swapping sides, and those that create their own substitutes. Then it needs to add the influence of factors such as ocean currents and the way they tip the playing field. Finally, it needs to be able to add the extent that humans have bolstered both Warms and Cools. After all that, once such a computer model can mimic past changes in climate, it can be used to predict future changes, based on varying assumptions about future changes in human activity.

The task may sound simple enough – but in reality it is virtually impossible. The plain truth is that there is not a computer model in existence that could swallow data on the climate of 1900 and give an accurate description of the climatic history of the twentieth century – certainly not without a great deal of tinkering using the benefit of hindsight. Everything is problematic. Even basic data on current world climate are sketchy and riddled with extrapolations. Sketchy, too, is an understanding of the way that solar radiations, greenhouse gases and the different layers of the atmosphere interact with each other. And most problematic of all – and yet probably absolutely crucial – is how to provide computer models with measures of such complicated three-dimensional variables as cloud formations and ocean currents. One example is the effect on global warming of El Niño, the erratic reversal of the warm currents in the Pacific Ocean. El Niño events can cause climatic chaos around the world, and are unpredictable at the best of times. If we couldn't predict the climatic influence of El Niño in the 1990s, how can we possibly expect to predict its influence in the 2090s?

It is not surprising, then, that all the preliminary attempts to understand the human contribution to changes in climate over the past century have been studded with arguments and contradictions. Only two things are really agreed. The first is that over the past century, the earth's temperature at the surface has risen by about 0.6 deg. The second is that over the same time period the level of carbon dioxide in the atmosphere has increased – largely, if not entirely, as a result of human industrial processes. Naturally, it has been tempting to blame the latter for the former – but there are many inconsistencies that need to be explained away before science can expect governments – and people – around the world to act on their conclusions with unflagging determination.

For a start, if the swing to the Warms of 0.6 deg. has been due to our bolstering of greenhouse gases, the warming should have been seen all over the world and through the whole 10 kilometres or so of the troposphere. Yet, for the past twenty years, while temperatures at ground level have been showing their most dramatic rise, the warming has failed to penetrate the atmosphere. Since 1979, long before global warming was an issue, satellites have been measuring temperatures in the troposphere. So, too, have weather balloons, data from which are archived in the British Meteorological Office. So far, and in stark contrast to ground-based meteorological stations, the satellites and balloons have picked up little evidence of warming. As we noted earlier, at heights above the earth as low as 3 kilometres, the temperature has been at least stable if not falling.

This inconsistency is important – and is just one among many.

For example, even at the surface the rise in temperature has not been as great as the best computer models would have predicted simply from the rate at which carbon dioxide has been increasing. They would have expected a temperature rise about twice as large as observed over the past 150 years – about 1 deg. rather than the observed 0.5 deg. to 0.6 deg. And for recent years, the models would have expected a swing to the Warms of three times what we have seen. Furthermore, the rise in temperature over the past century that triggered the prediction of global warming has been far patchier than it should have been. It has been evident only in the temperature data recorded at night and in winter. While night-time temperatures have risen by an average 0.8 deg. over the past century, those during the day have been virtually static and in many places have fallen: for example, maximum temperatures between June and November averaged over the land areas of the northern hemisphere for the past forty years show a cooling of 0.4 deg. Finally, the geography of warming and cooling is also a problem. According to global-warming models, the northern hemisphere should warm faster than the southern – because humans produce more greenhouse gases in the north. Yet, in the mid-twentieth century, the southern hemisphere was warming faster than the northern. The reverse has been the case since the late 1980s – but still not exactly as expected – because since 1987 the southern hemisphere has actually been cooling, a trend that no model based on human interference can yet explain.

Some scientists claim that inconsistencies such as these are enough to dispel the fears that there is a human hand behind the climate changes of the twentieth century. They claim that the global-warming models are simply wrong; that our support for the Warms via our production of greenhouse gases has been overestimated and is actually irrelevant to the outcome of the thermal tug-of-war. The most likely reason for error is that all computer models have assumed a positive feedback between different members of the Warms side. In particular, they assume that as carbon dioxide warms the world more water vapour evaporates into the atmosphere. And, as a greenhouse gas itself, water vapour works increasingly hand in hand with carbon dioxide to warm the world still further. In other words, the harder one Warm pulls, the harder another pulls. But positive feedback is far from proven. If, as satellite and weather-balloon data suggest, the free troposphere (above hills and mountains) is largely cut off from the surface, water that evaporates from the oceans will not necessarily mean more water vapour in the troposphere. Even if it does, as the clouds are warmer, maybe they will turn a greater proportion of their moisture into rain. Result: a drier free troposphere and a negative water-vapour feedback on carbon dioxide. In other words, as humans help one member of the Warms side to pull harder, another member eases off. So which is the more likely – positive feedback or negative? The IPCC in its *Second Assessment* published in 1996 admitted to serious gaps in the understanding of water vapour, conceding that feedback 'remains a substantial uncertainty in climate models'. Yet without a positive feedback, the human contribution to the thermal tug-of-war leads merely to a pleasant 1 deg. warming of the world, not a catastrophic 4.5 deg. warming. Global warming would then be a benign companion to humans through the twenty-first century rather than a menacing spectre.

However, there is an alternative explanation for the slower than expected rate of warming; an alternative that still leaves global warming as a real threat. Perhaps human reinforcement of the Warms is being masked more than expected by simultaneous support for the Cools. That although our production of carbon dioxide really is acting undesirably to warm the world, much of this warming is, for the moment, being masked by the sulphate smog that is also emanating from industrial areas. Many scientists are convinced that once their computer models get the balance right between human support for the two sides in the thermal tug-of-war, they will be able to describe past changes in climate very well and hence have a chance of predicting the future.

Their current calculations actually suggest that globally the warming effect of our production of greenhouse gases should be several times greater than the cooling effect of our production of sulphates. Therefore masking does not easily explain why the rise in temperature has not been faster. But there is much that is still unknown about aerosols and cannot be included easily in the models. Our support for the Cools from the burning of forests and other biomass is problematic. So, too, is the cooling effect of desert dust. The volume of dust entering the atmosphere dwarfs both sulphate and biomass aerosols, but the cooling effects are still largely unknown. Another problem is that greenhouse gases are long lived whereas sulphate aerosols survive only a few days in the air. So, while the effect of greenhouse gases is global, the shorter-lived aerosols act on a regional scale. To incorporate an average cooling for sulphates in models would therefore be fairly meaningless; their effect is so patchy. In most of the southern hemisphere, their impact is virtually nil. But in much of the northern hemisphere, summer aerosol cooling can locally offset greenhouse warming completely. Some estimates even suggest that local cooling, rather than global warming, may be the most immediate threat in the coming decades for the majority of the world's population – the whole of Europe, the eastern United States, half of Russia and most of China and the Far East. Current cooling due to aerosols over the eastern United States is more than twice the predicted warming due to greenhouse gases. Over eastern China it is three times, while over central Europe it is a staggering four times.

Clearly, models of the human role in the earth's climate are at an uneasy stage of uncertainty. It is just possible that total human support for the Cools is great enough to mask our support for the Warms, thereby explaining away the surprisingly small rise in global temperature over the past century. In which case, if we reduce our support for the Cools by cutting back on our emissions of sulphates in the interests of pollution control, the Warms would gallop away with the contest and our futures. However, if the positive feedback on which computer models depend are largely imaginary, the climate could simply be less sensitive to human activity than we were first warned. The only thing that is certain is that the current models are neither powerful enough, sophisticated enough, or informed enough to be able to decide.

How Stable the Sun?

Human activity is not the only possible cause of the warming observed at the earth's surface in recent decades. Another is that the sun is growing stronger. The sun is a star and as such it once was born and one day it will die, taking earth with it to its cataclysmic grave. In astronomical terms, the sun's days are numbered. Fortunately for us, though, solar middle age is energetic, stable and long lasting. Terminal senility shouldn't set in for about another 5,000 million years.

Stability, though, is always relative. The sun's energy output is not absolutely constant. It does vary a little, both in the short term in fairly regular eleven-year cycles, and in the long term. For example, the length of the solar cycle is currently steadily decreasing and is now several months shorter than it was at the end of the nineteenth century. Shorter solar cycles imply more solar activity, which in turn means that more solar radiation is reaching the earth now than a century ago. In short, the sun is growing stronger.

We could be excused for following intuition and assuming that a more active sun over the past century must have bolstered the Warms in Gaia's thermal tug-of-war. Unfortunately, intuition isn't enough – and as soon as we ask how much the sun could have tipped the balance, we hit a problem because the changes in solar activity have actually been very small. The energy emitted over a single eleven-year cycle varies by only a tiny 0.1 per cent – and from cycle to cycle, the differences are even smaller. So can we forget the sun in our effort to predict how the tug-of-war will play out in the future? Probably not – because it is possible that these tiny changes in total solar energy conceal much greater changes in other aspects of the sun's heating power.

Suppose it wasn't the total energy from the sun that mattered but only a part of it – a part that fluctuated much more than the total energy. Ultraviolet (UV) radiation, for example, fluctuates three times as strongly during a solar cycle as total radiation. UV light could set up a cascade of events from its interaction with ozone in the stratosphere. Alternatively, along with X-rays from the sun, UV could begin the cascade from even higher in the atmosphere: nearly 500 kilometres into space at the outer limits of the thermosphere where the temperature averages a scorching 1,500 °C. Both cascades are purely hypothetical, but in principle either could amplify changes in temperature at the earth's surface more than enough to matter.

A completely different possibility is worth mentioning if only to show how feeble links can be in the science of global warming. It involves a tenuous connection between solar wind, cosmic rays – and clouds! The more active the sun, the stronger is the solar wind – the charged particles streaming from the sun. The solar wind is known to push aside cosmic rays, preventing them from bombarding the earth – and cosmic rays perhaps encourage clouds to form in the earth's atmosphere. So the reasoning goes that the more active the sun, the stronger the solar wind, the fewer cosmic rays reach the earth's surface and the fewer clouds are formed, thereby reflecting less of the sun's energy back into space and leading to a warmer world.

So how do we decide which has played the greater role in warming the earth's surface over the past century – a stronger sun or human pollution? There is only one way; we are back to computer modelling. In one analysis, the best match with changes in global temperatures recorded over the past 200 years was obtained when all blame was attributed to the sun, and none to humans. But – such is the nature of models and statistics – the match was almost as good for the same period if all blame was placed on humans. As if this weren't even-handed enough, by changing the assumptions slightly it was possible to partition blame. For example, one conclusion was that a slow increase in solar activity was largely to blame for the 0.3 deg. rise in global temperatures between 1890 and 1970 but that human activity was to blame for the further 0.3 deg. warming since 1970. Another conclusion was that the solar influence on global warming between 1890 and 1997 was only 20 to 40 per cent.

Maybe the thesis of an increasingly active sun and complex human interference with both sides in the thermal tug-of-war really can explain the pattern of the past century's slow rise in global temperature. Alternatively, jaundiced observers could conclude that because a whole range of explanations can be claimed to fit the observed changes, there is as yet no real reason to accept any one of them. Yet the most pessimistic oracles insist that to ignore the warning signs now could commit our descendants to a dire environment that we could have averted. So . . .

Can We – and Should We – Change the Future?

If the changes in temperature over the past century, in all their complexity, were increasingly being driven by human activity then, in principle at least, we should be able to avoid future catastrophe by changing that activity. On the other hand, if an ever-more-active sun was driving those changes, we may still need to change our activity – but to counteract the sun rather than our own self-destructive proclivities.

Suppose for the moment that humans rather than the sun have driven the last century's climate changes – and that the only reason the rise has been so small is because so far our production of aerosols has masked our production of greenhouse gases. The fear is that one day the aerosol mask will fall away from the face of global warming and suddenly expose its underlying horror. But is such sudden unmasking really a danger? After all, it is largely the same range of industrial processes that produce both greenhouse gases and aerosols. If it really is us driving the changing climate, perhaps we can relax a little and assume that we are unlikely to reduce our production of aerosols without simultaneously reducing our production of greenhouse gases. Sadly, the answer is no. If we are driving, we can't relax. The timescales for braking and acceleration are entirely different. If we shut down all the world's power stations today, the cooling from their aerosols would virtually disappear within a week. But the greenhouse effect from past emissions of carbon dioxide would continue for many decades. The planet may already have reached the point where shutting down all the power stations would simply create a massive and entirely unpredictable convulsion in the atmosphere and its weather systems.

So if we are driving global warming, what should we do about it? Reduce the rate at which human-made greenhouse gases are released into the atmosphere, of course – but how? Among the most colourful suggestions are either to remove carbon dioxide from smokestacks and stash it underground or to fertilize the oceans to increase their uptake of the carbon dioxide that we do produce. But by far the safest and most sensible solution would be to cut down on the burning of fossil fuels by cars, power plants and other major users.

All such sensible actions, of course, are probably pointless if it turns out that we are not driving climate change. Suppose that, after all the hype about our destruction of the earth's thermostat, it transpires that the real culprit is an ever-more-active sun; that our contribution to the balance of power in the thermal tug-of-war is actually minuscule. This would not make global warming any less of a threat – or the consequences any less dire – but it would require a completely different response. Reducing our support for the Warms by cutting back on carbon-dioxide emissions would still help, of course, but priority would need to switch to bolstering the Cools as much as possible. What could we do? Shield the earth from the sun with mirrors of aluminium foil orbiting in space? Increase the reflectiveness of the earth by deliberately dumping large amounts of sulphur dioxide in the stratosphere? Or would that destroy the ozone layer and make matters worse rather than better?

And herein lies the problem; the big question. Do we really understand what is happening to our climate – and why it is happening – well enough to risk doing anything? Which is worse, to do nothing for the moment, or to at least do something and risk making matters worse rather than better? Well, the decision has already been made. Things are being done. Or, at least, they seem to be.

So What Is Being Done? – and Hands Up Those Who Want It

The first warning that industrial activity would eventually have an impact on global climate came over a hundred years ago, in 1896, from a Swedish chemist, Svante Arrhenius. No action was taken, though, until the modern vision of global warming first hit the front pages in 1988. Near panic ensued and in the same year the UN set up the IPCC. Since then, there has certainly been a lot of committee and panel formation – and a lot of talking by a lot of people. In fact, few environmental problems have received such attention at such a high and international governmental level – not least at the Earth Summit in Rio in 1992. Since then, many governments around the developed world have set about reducing their output of greenhouse gases.

Measures to combat global warming, though, are not welcomed by everybody, even in the developed world. Whole industries – and maybe whole countries – stand to suffer economically if the thermal threat is opposed with too much dedication. A new category of computer model has entered the arena: the economic model. This is the weapon of choice for those who oppose any reduction in the emission of greenhouse gases. Various assumptions are made that are even more tenuous than the science. For example, some models claim that in economic terms the cost of a 3 deg. global warming would only be 1 or 2 per cent of today's gross world production. How can anybody possibly calculate such a figure – given such uncertainties in the underlying science? Not surprisingly, though, with such a small assumed cost, models usually show that economically it is best to do nothing at all about global warming; we should continue to pollute the atmosphere.

Whether a person is concerned that something will – or won't – be done about global warming, there are enough potential traumas swirling around within the different scenarios to worry everybody. And in the late 1980s and early 1990s, nearly everybody was worried – hence the establishment of the IPCC and the calling of the Rio summit. The IPCC is still in existence; still beating its collective heads against the combined walls of scientific discord, governmental reticence, industrial opposition and – in recent years – general disinterest.

A new phrase has entered the global-warming dictionary – 'greenhouse fatigue'. With the possible exception of those scientists whose lives and careers have come to depend on making global warming seem vibrant, exciting and threatening, everybody from governments and industry to the person in the street is becoming increasingly bored by the whole business. Even the media – at the dawn of a supposedly apocalyptic new century – now struggle to find anything of dramatic interest to say about the future of the earth's climate.

Prudent Inertia

Greenhouse fatigue was inevitable. Despite the early hype, the early panic and the rapid government attention at the highest level – and despite the applied minds of the best scientists in the field – the truth is that we are no nearer the truth about global warming than we were in 1988. Maybe the earth is warming up to an extent that matters, but maybe not. Maybe humans are driving climate change, but maybe not. Maybe we can change the course of climate's evolution over the next century, but maybe not. Scientists haven't yet found enough consistent evidence or powerful enough analytical tools to reach a consensus even among themselves, let alone to convince self-interested governments and industries. There are two very good reasons for science's impotence over the matter: first, the situation is impossibly complex, and secondly, it defies proper scientific study.

There cannot be experiments – or controls. All analyses and interpretations depend on correlation – or correlations of correlations – or even correlations of correlations of correlations, and so on. The science is further removed from its target than anything we have discussed so far.

Of course, as far as the scientists and sciences associated with global warming are concerned, the last two decades of feverish interest – and funding – have been invaluable. Many careers have flourished, a great deal has been learned and the scientific study of global climate has advanced enormously. One day, the things that have been learned during the impetus of the past twenty years or so will contribute to a solid understanding of the subject. But for the moment the problem remains beyond robust study. No prediction – whether apocalyptic or reassuring – can be anything more than intuition and guesswork. Somebody will turn out to be right, of course. Among all the attempts, somebody is bound to have forecast the future climate correctly. But that person's success is as likely to be chance as it is to be science. After all, if enough people read enough tea leaves, one of them would predict the future accurately too.

So if the science of global warming is unable to help, what should the governments of the developed world do in the face of the global concern generated by a few scientists? Should they do nothing and risk being accused of not caring? Or should they do something and risk making mistakes? After all, dramatic actions such as shutting down all power stations are as likely to be catastrophic as they are to be inspired. Frustratingly, the most prudent course of action is probably the one least likely to win votes: to wait and see. They, and we, have to accept that, for the moment, we simply cannot tell whether our great-great-grand-children will live in a climate very different from ours. We cannot tell whether they will thank us or blame us. But the chances are that if we could look into the future and see their faces, they would simply smile back at us in wry amusement at how easily we could be panicked by scientific hype.

6

Conservation:
A War of Two Worlds

The world's plants and animals are disappearing. Not since an asteroid slammed into the earth around 65 million years ago have species died out so fast. This time, though, the fault is ours. Humans are recklessly clearing, polluting and killing the habitats that are home to the products of millions of years of evolution. Nearly half of the world's rainforest has already been destroyed and a further area the size of Florida is lost each year. These forests are home to more than half the world's species of plants and animals. The statistics are staggering: 1 hectare of forest in Brazil yielded 425 kinds of trees; one corner of a Peruvian National Park contained 1,300 species of butterfly. To the human denizens of impoverished city habitats, such sparkling diversity is beyond experience – almost beyond imagination – and is about to vanish. Species are disappearing up to 1,000 times faster than before the coming of humankind. Nearly 1,000 species of tree are on the edge of extinction, the giant panda now numbers fewer than 1,000 individuals and the Javan rhinoceros numbers fewer than 100.

The catalogue of environmental disasters goes on – so clearly we must stop now before it is too late. Forests buffer the earth's climate against unwanted climatic change; they are also huge food stores and apothecaries, providing us with a multitude of medicines. Wanton destruction of habitats will rob Third World countries of a natural resource that could earn them valuable income from that new breed of tourist, the eco-tourist. And, of course, if we don't act immediately and forcefully, future generations will inherit an impoverished and homogenized world, robbed of the chance to see nature's wonders at first-hand.

This is the hype – and many of us were exposed to it from an early age. Few adults in the developed world would even begin to question the conservation ethic. But how much of it is science, how much aesthetics and how much hypocrisy?

The Seventh Extinction

Nobody knows how many species of plants, animals and other organisms there were on planet earth at the dawn of the twenty-first century. Biologists have so far described a total of between 1.5 million and 1.8 million species – but this number is just the known tip of an unknown iceberg. Some estimates put the number waiting to be described as high as 100 million, though 10 million is perhaps more realistic. Mainly it is the smaller organisms that are waiting to be discovered: perhaps 90 per cent of insects, nematodes and fungi and as many as 99 per cent of bacteria and other micro-organisms.

Any list of existing species is merely a snapshot in time – because organisms are forever evolving and dying out. The 'birth' and extinction of species has been going on since life began. It is a natural process. Probably 99 per cent of species that have ever existed over the past 600 million years are now gone, their average lifespan perhaps a mere million years. Long before the industrial revolution, humans had seen many changes. Our Kenyan ancestors nearly 2 million years ago mingled on the rich grasslands and forest fringes with animals that no longer exist. There were short-necked giraffes with huge antlers, antelopes as large as buffalo with massive curving horns, carnivores which looked like sabre-toothed lions, and many more.

Most of the time, the rates of appearance and disappearance of species are about equal so that the total number of species on the planet stays roughly constant. From time to time, though, something causes a large proportion of species to die out. Buried in the fossil record of the past half-billion years, palaeontologists claim to have found six such mass extinctions. The most recent and most famous happened 65 million years ago at the end of the age of dinosaurs. It was probably caused by a giant meteorite smashing into the earth off the present-day coast of Yucatán, thereby changing the planet's climate and chemistry.

Past mass extinctions and recoveries have followed a similar sequence. First, a large proportion of the pre-existing plants and animals die out. Soon, a small number of species – mainly fungi and ferns – take advantage of the space emptied of other life and rapidly spread. They are joined later by a few species that managed to cling to life in isolated pockets of habitat and eventually re-colonize locations from which they had been wiped out. Then, very slowly, across 2 to 5 million or more years, life as a whole evolves back to its full, original variety, but with species different from those that had existed before. If that meteorite hadn't collided with earth 65 million years ago, mammals may never have spread as they did and we humans may never have evolved.

Biologists claim that we are currently in the middle of a seventh mass extinction – and this time the fault is ours. We have, of course, hunted a few species to extinction – mammoths, dodos and passenger pigeons to name a few – but this is not our main modus operandi. Our most powerful exterminations ride on the destruction of habitats on which many species depend to survive. Of chief concern at present is our destruction of tropical forests.

Forests are one of the planet's critical natural resources. They contain half of the world's species, act as a 'sink' for atmospheric carbon dioxide, and provide a sustainable source for many forest products, from timber to fruits, traditional medicines and meat. Yet each year some 170,000 square kilometres of tropical forests are lost. Some nations are destroying their trees at an annual rate of 5 per cent. Others are replacing natural forests with monoculture plantations – of eucalyptus, for instance – which despite holding carbon dioxide and yielding timber provide few other forest products and cannot support traditional forest communities. And as the forests disappear, so do the wildlife they contain. The symbol of the World Wide Fund for Nature (WWF) is the giant panda – because there are now only around 1,000 wild-living individuals scattered among the receding bamboo forests of China. The largest surviving group numbers just seventy individuals. With the bamboo forests destined to shrink further – and despite the establishment of reserves – some biologists predict the giant panda will be extinct within a decade.

Forests and their inhabitants are not our only victims. Half of the world's wetlands have been lost in the past century. Around two-thirds of all coral reefs are imperilled by human activity, 80 per cent of grasslands are suffering from soil degradation, half of all salt marshes and mangrove swamps have been eliminated or altered, 20 per cent of dry lands are in danger of becoming deserts, and groundwater is being depleted everywhere. The steeply sloping sides of mountain regions are suffering soil erosion and forest loss. In mountain ranges from the Himalayas to the Andes and the Alps knock-on effects are landslides and altered river flows. Finally, land-based pollution is a major contaminant of the oceans and overfishing is decimating stocks of fish and other marine animals. As a result of all this destruction, around one-fifth of freshwater species have vanished or been driven towards extinction in recent decades and about a quarter of known plants are expected to face extinction by the year 2050. If the current rate of habitat destruction continues, half the known species of plants and animals will be gone by the end of the twenty-first century.

It is, of course, possible to quibble with these figures. They are crude, anyway, and all assume that the species we don't know are disappearing at the same rate as those we do know. This may or may not be true. There are also value judgements involved in our responses: we mourn the dodo but not the smallpox virus. But give or take a few per cent, the predictions being made by conservation science are uncompromising. If nothing is done to halt the seventh extinction, our descendants will inherit a world with few life forms and an environment that will look much the same over large parts of the planet.

Global Warfare

It is often said that humankind is destroying the planet and that the problem affects us all. It's an exaggeration, of course. Even a nuclear holocaust would not destroy the planet. The phoenix of biodiversity would rise from the ashes, just as it rose from the ashes of the previous mass extinctions. Ecosystems would evolve again, even if based on weird and wonderful new assortments of plants and animals. The planet will continue until an expanding sun finally swallows it in 5,000 million years' time. Our actions are not killing the planet – they are not even making it uninhabitable – but they are certainly changing it. The question we have to ask is whether that change is for better or for worse – or whether it is simply change. After all, was the last mass extinction 65 million years ago a good event, a bad event – or just an event? Scientifically, we cannot blindly assume that conservation is 'a good thing'; first we need clear evidence of a universal benefit from saving planet earth from a seventh extinction.

To people raised amidst the conservation hype, even questioning the ethic seems like blasphemy. If any scientific discovery should bind humankind in common endeavour it is the threat of a mass extinction. It comes as a surprise, therefore, to discover that the scientific arguments for 'saving the world' do not impress everybody. On the contrary, in places they are a source of anger and conflict. Most of the disagreements surfaced forcefully during the so-called Earth Summit convened in Rio de Janeiro in 1992. The avowed aim was to bring together the governments of the world to steer the planet safely through the obstacle course scientists foresaw during the coming century. In the end, all it really did was highlight the deep divisions between nations over the real problems and acceptable solutions. These divisions still exist and show no sign of going away, whatever warnings scientists continue to give.

The main cause of the division is that the pressure for conservation emanates primarily from the developed world – but is directed largely at the Third World. This is because much of the world's remaining biodiversity resides in its poorest nations, especially in Asia, Africa and Latin America – and many such nations view the conservation movement as political, subversive and hypocritical. They consider the science of conservation to be not only weak but also politically biased; environmental protection is just another instrument for the rich nations to dictate to the poor.

A major disagreement is over global warming. As we saw in the previous chapter, scientists from the developed world are arguing that global warming is for real. And by and large, despite the weaknesses in the science itself, they have managed to convince their governments and public alike that they are right. As a result, the developed nations are promoting global warming as a global problem over which all nations should cooperate. Hence, the Third World should stop destroying its rainforests so that the planet can continue to mop up the huge amounts of carbon dioxide being produced by industrial processes. In contrast, Third World nations see that the science of global warming is weak; but consider that if the predictions do have substance, any problems have been created by the developed world and will be suffered by the developed world. Solving those problems should therefore be of little concern to the Third World.

Another disagreement centres on the developed world's wish to maintain biological diversity (biodiversity) so as to halt the seventh extinction – which can also only be achieved if Third World governments take heed of what they are told by developed world scientists. The Third World is not impressed. They want the same rights that developed nations have always enjoyed: the right to develop and exploit their national resources in the way they think is best for themselves and their people.

Forest Histories, Forest Futures

There is some difference between what scientists calculate is needed to combat the twin dangers of global warming and reduced biodiversity: the former is needier – or greedier – of Third World resources than the latter. Forest destruction has to stop now to minimize global warming whereas far less of the earth's surface is needed to save a large proportion of threatened species. Conservationists point out that there are hot spots that contain high concentrations of endangered species. From the coastal sage of California to the rainforests of West Africa, the hottest of terrestrial spots occupy only 1.4 per cent of the earth's surface yet are the exclusive home of more than a third of the terrestrial plant and vertebrate species. Similarly, from the streams of Appalachia to the Philippine coral reefs, aquatic hot spots occupy a tiny fraction of the shallow water surface.

It will be pointless, though, conserving only small pockets of habitat. Plants and animals live a precarious existence when their habitat is too fragmented. Usually, they need untouched areas of a reasonable size if they are to maintain themselves, and often they need 'corridors' of vegetation permitting free movement from place to place. If one species dies out through lack of space, so do others. When a single plant species becomes extinct, thirty other species – largely pathogens – become extinct with it. If an animal becomes extinct, so may its main predators. Biologists may disagree over whether complex ecosystems are more or less fragile than simple ecosystems, but whatever the answer the vast majority of plant and animal species cannot survive outside of large unaltered chunks of the habitat for which they evolved.

Essentially, therefore, conservation scientists are saying that global warming and mass extinction can only realistically be averted by rapid and positive action by the Third World: relatively large areas of forest and other habitats need to be set aside now and conserved. Human encroachment and perhaps even access will need to be forbidden. Only in this way can the world's production of carbon dioxide be mopped up; only in this way can valuable plants and the largest animals be assured of survival.

It could have been just coincidence that most of the world's biodiversity is to be found in the poorest nations – but it isn't. The explanation is partly ecological – diversity will always be greatest in the tropics and least at the poles because of gradients in solar energy – and partly historical. The developed nations actually owe their rich status to a long history of destroying their forests and other habitats. They have relentlessly exploited natural resources as a way of simultaneously obtaining raw materials and making room for people, agriculture, industry and cities. Deforestation made the developed countries what they are today; industrialization and national wealth were financed by the destruction of the northern temperate forests and the pollution of rivers, lakes and oceans.

Forests did not always cover the land. Twenty thousand years ago, at the peak of the last ice age, Europe and North America were covered by ice and tundra, not trees. Only the temperate forests of Asia were relatively unaffected. Even the Amazon basin was forested only in patches, not by the dense cloak we know now. But when the world emerged from the ice age around 10,000 years ago, the forests spread and at their peak covered 60 per cent of the land's surface. Since then – with more people clearing more forests to make room for more crops and livestock – the forests have dwindled. Now they cover only around 25 per cent of the land – with most of the devastation having occurred on the territories of the now-developed nations.

Most of this developed world deforestation took place in the few hundred years leading up to 1900. Since then, the process has slowed down or stopped. Thanks to replanting schemes, Europe and North America can boast that they have ended the second millennium with as many trees as they had 100 years earlier. These forests, though, are no more than a ghost of those that existed at the end of the first millennium. And alongside the demise of the habitats went the demise of wildlife. Americans had exterminated their passenger pigeon and the British their wolves, bears and boars long before the Chinese began to threaten their giant panda, Javans their rhinoceros or Africans their mountain gorilla. The Third World still has some way to go before it has destroyed as many trees, habitats and large animals as the developed world. Even those – such as the dodo – that have disappeared from the tropics were often killed by itinerants from the developed world rather than by indigenous peoples.

Habitat destruction in the developed world is still far from over. The Aral Sea, a huge freshwater lake in Central Asia, has shrunk to under half its former size after its two main rivers were dammed to irrigate cotton fields in Uzbekistan. 'Let the Aral Sea die a beautiful death. It is useless,' proclaimed a Soviet planner in 1987. Since 1960, England has built on land equal to the size of Wales. In parts of the United States – Ohio, for example – land 'development' around urban areas is growing five times faster than the population. Only Tokyo is no longer expanding – because it has filled the surrounding Kanto Plain and so has nowhere further to go.

Most urban encroachment, of course, is not at the expense of untouched habitats – because there aren't any. Instead, it is at the expense of agricultural land. Since 1981 the amount of land worldwide devoted to raising grain has fallen by 7 per cent. So far, this loss has been compensated by increased agricultural efficiency and productivity, but a limit will soon be reached unless genetic modification of crops – discussed in the next chapter – can raise the ceiling of productivity.

Even when a developed nation tries to halt the process of habitat destruction, it is difficult and expensive. In part, this simply illustrates that it is easier not to destroy habitats in the first place than it is to restore them – but it also reflects the difficulty of marrying conservation with commercial and domestic pressures. The United States, for example, is struggling to halt the decline of the Everglades, the river of grass that once covered 11,700 square kilometres in Florida. Over the past century, the channelling, damming and diverting of water for urban and agricultural use have caused the ecosystem to collapse. Nearly 70 per cent of resident species, including manatee and panther, have become endangered and the 'glades' capacity to store water has shrunk as human demand grows. With Florida's water supply and a $14 billion annual tourist industry in jeopardy, a 1998 plan to reverse the process began, including removing phosphorus from agricultural run-off. But there is no guarantee that the project will succeed; the prospect looms that even the world's richest nation is unable to reverse the process of habitat destruction – perhaps because, in reality, it is unwilling to deprive the local population for the sake of the environment.

Not surprisingly, therefore, the poorer nations are unimpressed by what they see as hypocritical pressure to stop clearing their forests to make room for their people and to fund their development. The Third World sees their forests as a resource, to be used by their people in the most profitable way just as the developed world used – and still uses – theirs in the most profitable way. And, at present, profit means selling the timber and using the land for agriculture. Their counter to the scientists who predict global warming if deforestation does not stop is to suggest that the developed nations should stop producing carbon dioxide and regrow their own forests; let the developed nations feel the pressure of balancing people and environment for a change and see how much notice they take of conservation science then.

Overpopulation

One reason that humans destroy habitats is to feed and make room for their ever-increasing population – and the faster the population increase, the faster the habitat destruction. This is what drove northern nations to destroy their forests over the past few hundred years – and is driving southern nations now. The global human population has been increasing ever since the species evolved and at an ever-increasing rate. The rate finally peaked in 1965 at 2.04 per cent per year. Then the increase began to slow. Now it is only 1.3 per cent per year and still declining.

Current estimates are that from its present level of 6 billion, the human population will reach around 10 billion by 2050, and should level off at about 11 billion by the end of the century. This is good news – but means that a lot more mouths still need to be fed and a lot more living space still needs to be found over the next hundred years. Exactly how many more mouths and how much more space depend on what happens to human fecundity in the future and how this differs between developed and developing nations.

Just over 200 years ago, there was actually little difference in birth rates between the now developed and now Third World countries – each woman gave birth on average to about seven live babies in her lifetime. Then, in northern nations first death rates and later birth rates began to fall, beginning first in France in the late 1700s. This trend – the so-called demographic transition – has continued over the past two centuries and in most developed nations fertility now averages around or below two live births per woman. In some places – Spain, Slovenia, Greece and Germany – it is below 1.5. The same decline did not begin in Third World countries until much later. A few decades ago the figure was still 4.9 babies per woman and now is 2.7, but in parts of the Middle East and Africa it remains around seven. This difference between developed and Third World nations has led to differences in the pressure on space and agriculture. For example, in 1950 Europe had three times the population of Africa. Now, the two are about equal – and by 2050 Africa will have three times the population of Europe. Currently, the Third World is accounting for 96 per cent of the annual increase in global population. Consequently, most of the pressure to find more space and feed more mouths is being felt in the Third World.

If populations were not increasing so fast in the tropics, there would be less pressure on the indigenous people to clear habitats in order to feed their families. Conservation science therefore places high priority on controlling the rate of population increase in the Third World. However, both the arguments and the usual solutions make an assumption: that the Third and developed

Worlds' fertility differ only because of differences in education (on population and birth control) and the availability of modern contraceptives. Hence the first step towards global habitat conservation is improved education and more freely available contraception in the Third World. There is a hint of arrogance here – the developed world congratulating itself on controlling its own population increase – that may not be scientifically justified. The evidence that fertility in the developed world declined because of better education and increased availability of modern contraception is based purely on correlation. Cause and effect has not been established. Undoubtedly, such modern developments have played some part in changing fertility but they are certainly not the only factors and may not even be the most important.

Fertility rates began to decline in Europe long before either people feared overpopulation or modern contraceptives were available. Differences between nations do not neatly relate to differences in contraceptive use. Some of the lowest fertility rates in Europe – such as in Spain – are found where modern contraceptives are least used. Nor do they relate to efficiency of contraception. Births to Japanese women showed the same decline as births in other industrial countries despite the contraceptive pill – the most efficient contraceptive – being banned until 1999. There has to be another factor – and the most likely is life expectancy.

Throughout human history, women have given birth to the number of children that will give them two to three who survive to maturity. At times and in places when survival prospects were good, women had fewer children, on average, than at times and in places that prospects were poor. There is as good, if not better, a correlation – both historically and geographically – between fertility and life expectancy as there is between fertility and the use of modern contraceptives. When a woman is subconsciously fairly certain that each child she bears will survive to maturity, her psychology and physiology automatically reduces the number of children she conceives. If modern contraceptives are to hand, she will use them – but even if they are not she will not have endless children. In the early 1900s, without a contraceptive in sight, hunter-gatherer women, whose children had an excellent chance of survival, used to give birth, on average, to only three to four children. In contrast, when a woman fears that every child born is likely to die, she will have many – as in developed countries two centuries ago and as in many Third World countries now. Even if modern contraceptives are to hand, a woman will not use them assiduously if she thinks most of her children are likely to die.

The last major World Population Conference in Cairo in 1994 concluded that the best contraceptive was affluence and security. In modern human populations, childhood survival depends on wealth. Even in a developed country like Britain, a child born into a low socio-economic group has twice the chance of dying during childhood than one born into a high socio-economic group. Wealth means health, both at the individual level via diet and lifestyle and at the national level via medical services. The higher level of fertility in the Third World may owe more to their lack of wealth and therefore health than to their lack of education or contraception. The implication is that not until childhood survival in the Third World rises to the same level enjoyed by developed nations will fertility rate also drop to the same level. In a sense, then, the reduced fertility of developed nations is the result of deforestation leading to wealth. The Third World may well have to follow the same road to population control.

Overconsumption

A reduction in population growth is not the only target in the developed world's push for conservation in the Third World. Another is what the former perceives as a lack of environmental policy by the latter: as well as the destruction of rainforests from the Amazon to Borneo they cite the black smoke pouring from the factory chimneys of India and China, the smogs of Mexico City, and the African irrigation projects that dry out more fields than they water. The inefficiencies of the Nigerian oil industry are so great that enough natural gas is flared off from oilfields to serve the entire energy needs of most of Africa. Between 1990 and 1995, China, India, Brazil and Indonesia increased carbon-dioxide emissions by more than 20 per cent; Turkey, Korea and Mexico are expected to have per-capita emissions as high as Europe by 2010. Against these environmental deteriorations in the Third World, developed nations cite their own improvements. They point out that their energy efficiency has improved by more than 20 per cent in the past two decades. Despite economic growth, carbon-dioxide emissions in the United States were less in the late 1980s than they were in the late 1970s. And better pollution controls have brought spectacular reductions in river pollution and acid rain. The developed world claims that, if current trends continue, the contribution of Third World countries to global emissions will approach 50 per cent by 2010.

The Third World, though, has a different perspective on such figures: even equal gross contributions to emissions in the future will not be seen as parity. The main complaint is what the Third World sees as the developed world's overconsumption. There are several ways of expressing this complaint. The richest 20 per cent of the world's population currently consumes 80 per cent of its resources and produces 80 per cent of its pollution. New York alone uses more electricity than the whole of Africa from the Sahara to the Limpopo. The average American contributes eight times more to the greenhouse effect than the average Chinese and twenty times more than the average Indian. Nine out of every ten children in the future may well be born in the Third World – but those nine children together will still consume less than one-third as much of the world's resources as that single child born in the developed world. However overconsumption by the developed world is expressed, the Third World sees it as a bigger problem than their own environmental policies. It will be many decades – if ever – before a child in the Third World consumes as much of the planet's resources as a child from the developed world. The onus, says the Third World, is on developed countries to put their own environmental house in order, not to bludgeon the Third World into doing it for them using the hammer of conservation science.

Attitudes might change, of course, if the Third World ever became convinced that global warming was both real and relevant: if the consequences were to be as dire for the tropics as for temperate latitudes. But the two main threats predicted for the tropics by global-warming scientists – desertification and the loss of coastal agricultural land due to rising sea levels – are both unconvincing. The loss of coastal agriculture would probably cost most Third World countries far less than conserving their forests. And most 1990s studies of desertification, such as from the analysis of satellite images, concluded that deserts had not spread during the twentieth century. Sands had grown and shrunk in response to annual changes in rainfall, but there was absolutely no sign of the year-by-year spread expected from global warming. To the Third World, the threats still seem to be motivated more by politics than by solid science.

There is real suspicion in the tropics that rich nations are not yet doing everything they can to react to their own predictions because in reality they are hoping to make the Third World pick up most of the bill. The 1992 refusal of the United States to sign the Rio agreement to reduce carbon-dioxide emissions by 5 per cent added to this suspicion – after all, thirty-four powerful American senators were in states producing coal and oil. Eight years on – at The Hague in 2000 – the United States would still not agree, even this time to limit an increase in emissions by 2010 to 7 per cent. The clear impression is that developed world governments are unwilling to pay in full for their global pollution – unless even they don't really believe their scientists.

Short Term, Long Term: Profitable Compromise?

A key element in the different perspectives of developed and Third Worlds – at least publicly – is timescale. The developed world, with relative security in the short term, can afford to show some concern for long-term threats to prosperity. The Third World is much more concerned with short-term development. As a result, the developed world accuses the Third of being short-sighted: of putting short-term profit and convenience before long-term planning for the planet's future. The Third World responds by saying that long-term planning is an indulgence that they can only contemplate once the short term is secure. People in absolute poverty will ravage their natural environment to try to survive, whatever the consequences for future generations. No one can be expected to fret about the kind of environment they will bequeath their great-grandchildren when they may not survive to produce any.

Accordingly, the Third World's attitude to global warming is simple. If the developed nations wish to use Third World forests to mop up their carbon dioxide for the foreseeable future, then they must pay a realistic price – with a profit margin – to have the service maintained. Of course, even this may not halt defor-estation. Just because a Third World government might see conservation as the most profitable use of their natural resources, the people actually living in and around the forests may not. Unless the benefits clearly trickle down to grass-roots level, habitat destruction will continue even if it is made illegal.

Developed world scientists will only get their way over global warming in the long term if forest conservation can be made a profitable option in the short term for those who would otherwise pay the real price. The same principle applies to the preservation of biological diversity. Nations with much of the population living below the poverty line are unlikely to be impressed by the developed world's willingness to spend large sums of money to secure the future of an elephant or orchid unless the people also receive some benefit. Not only the governments of the Third World but also the farmers and poachers will need to see more profit from species conservation than from killing and destruction – otherwise it simply will not happen.

There are two main ways in which such profitable compromises could be forged – via the interests of tourists and via the interests of pharmacological companies – but both compromises have their problems.

Eco-Tourists

One of the most-repeated reasons given in the developed world for the importance of conservation is emotive and aesthetic rather than scientific: we should preserve natural habitats for 'our children's sake'. Part of the argument is that if we destroy natural variety and turn the whole planet into a drab grey and green mosaic of concrete and agriculture we shall rob future generations of the chance to observe wildlife at first hand. The world would be a much less interesting place in which to live. There would be no more sights of tigers, elephants or giant pandas living the life for which evolution shaped them. Our children would never know the sense of wonder to be gained from watching wild animals in their natural habitat.

The problem with this argument is that in reality, first-hand experience of exotic habitats and species is limited to the wealthier echelons of humanity. It is not a universal benefit. Only people who trot the globe have cause to worry that one place looks very much like another. The majority will neither know nor care: even in the developed world, most people never travel to far-flung places. Their only contact with exotic wildlife is via TV screens, books or zoos – and these avenues can continue long after wild populations become extinct. And in the Third World, short-term survival is far more important than any potential disappointment from flying round a homogenized planet. Will enough change in the future to encourage poorer countries – the custodians of nature – to set aside large tracts of their land simply to give rich people from elsewhere a first-hand taste of their world? The only avenue to such international sympathy is if rich people pay for such a privilege – which they might.

Currently, around 600 million tourists visit foreign shores every year. Allowing for those who travel often, this is equivalent to about one-fifteenth of the world's population. Most, though, have little interest in world ecology. They travel in search of sun, sea and sand – and perhaps the experience of foreign culture – or they travel simply to visit far-flung relatives. Few are seeking to wade through warm mud while being soaked by torrential rain and attacked by snakes and leeches in order to experience life in a rainforest. Few, either, wish to spend an insect-bitten week waiting for the fleeting glimpse of a wild tiger when they can see one in comfort in a zoo. Even so, a new breed of tourist – the eco-tourist – is emerging; people for whom discomfort is a small price to pay for the first-hand experience of wildlife and wild places. They are a small proportion of the tourist market – at present no more than 1 per cent – but the recent proliferation of specialist tour companies suggests that this type of travel is a growth area.

Perhaps our children – eco-tourists of the future – will travel in their hordes to far-flung places to see the wonders of the natural world for themselves. If they do, then obviously they will pay their Third World hosts for the privilege – but will they pay enough? If they don't – if the local people cannot make more money from tourism than from selling or cultivating their land – there will very quickly be nothing left to see and nowhere untouched left to visit.

As far as conservation is concerned, there are two dangers from relying on visitors to support exotic habitats. The first is that tourism – even eco-tourism – might destroy the environment just as surely as if it were razed and cultivated. To ecological purists, there is no such thing as 'environmental' tourism. Just by travelling – most likely by jet – the tourist will contribute to polluting the earth, and on arrival they will damage the desti-nation. The only way to avoid adding to habitat destruction is to stay at home. Contamination by tourists is causing concern even in the Antarctic and is a problem in many other regions, from tropical forests to remote mountains. Wherever eco-tourists ven-ture in search of the ultimate 'unspoiled' destination, some form of development inevitably follows.

A prime example is the Mexican island of Cancún – a 10-mile stretch of sand that is currently the scene of one of the biggest mass migrations on earth. Until the early 1970s, the island had a population – descendants of the Mayas – of less than a hundred. They made their living by fishing and by gathering food from the scrub and forests on the mainland shore. The earliest visitors were undoubtedly captivated by its isolation and untouched beauty. Now, more than a million people fly in to laze in Cancún's tropical sunshine and the former 'island paradise' has become bedecked with some of the largest and most luxurious hotels in the Americas. Realizing its potential, the Mexican government and private enterprise deliberately transformed the place to earn foreign exchange. They succeeded – and the environment paid the price. Over the past thirty years, much of the island's original plant and animal life has disappeared, and the presence of so many visitors creates an enormous amount of waterborne waste that has to be absorbed by the shallow Caribbean coast.

The second danger is that not enough people will be interested in visiting an unchanged place to make eco-tourism sustainable. Some people will want to visit, of course: wherever wildlife exists in a relatively undisturbed state, from the tropics to the poles, some 'environmentally aware' tourists will be willing to pay large sums to see it. Already, eco-tourists not only pay more than the traditional sun-and-sand tourist, they also ask less. They are more inclined to accept a limited range of home comforts, and to adapt to local conditions. But will there ever be enough of them to sustain the habitats they visit?

Costa Rica, for example, is an ideal destination for the eco-tourist. Its small size, extraordinary biodiversity and stable political climate make it uniquely equipped to lure the environmentally curious. If eco-tourism cannot succeed in Costa Rica, its success elsewhere is doubtful. In the mid-1960s, the government set up a system of national parks, protecting about a tenth of the country's land area. Yet, despite the fact that three-quarters of foreign tourists visit at least one national park during their stay, the income this generates – just over half a million dollars – is nowhere near enough to maintain the park network. At each presidential election the future of the parks, and of tourism in general, become major issues. The future of even the most popular reserve is precarious: locals still have every incentive to fell more forest and graze more cattle.

Part of the problem everywhere is that the income generated by tourism often bypasses rural people and ends up in the pockets of city-based businesses and governments. In Cancún, most of the money ends up in the pockets of investors. Enough, though, reaches the people whose local environment has been destroyed for the majority to feel that suffering tourism is better than being unemployed. In Costa Rica, well under half of the tourist-generated income reaches the local people. For the moment, this is just about enough to quell resentment – but local sympathy for conservation is on a knife-edge. In East Africa and India, where eco-tourism is long established, the visitors produce very little revenue for local people as a whole. As a result, instead of welcoming eco-tourists, local people view them as being indirectly responsible for the theft of valuable land.

Eco-tourism is already expensive – beyond the reach of most people in even developed countries. Excluding the cost of airfares, the cheapest tours – such as a sixteen-day gorilla watch in Uganda or two weeks in the rainforests of Borneo – cost nearly $2,000. Slightly more expensive, both a fourteen-day bird-watching tour of Costa Rica and a seven-night cruise of the Galapagos Islands cost around $3,000. And a twenty-day cruise of the Antarctic Peninsula will cost up to $12,000. Yet still – in those destinations where people could live off the land in other ways – such prices aren't enough to sustain conservation and pacify all of the local inhabitants. Subsidies – from local governments or the WWF – are still needed. Most tours need to be more expensive still – and more of the money needs to find its way to the people whose cooperation is required if the destinations are to remain unchanged. Habitats that depend on tourism live too precarious an existence to avert the seventh extinction.

Food Stores and Apothecaries

Science has, though, produced a rather more compelling reason for conserving biodiversity than simply making life interesting for some of our children. The rainforests of the world are not only a source of food, they are also a rich and largely untapped medicine chest. If they are destroyed, humankind will be robbed of many vital drugs and medicines. The chances of conservation success, though, again depend on financial and other understandings between Third and developed worlds.

Plants already provide us with a quarter of our medicines and they could provide many more. Although medical practice in the developed countries over the past century moved away from dependence on plants for a while, in recent years there has been a revival of interest. The main focus is on the wealth of potentially useful species in the tropical rainforests. For example, the American National Cancer Institute has now identified around 3,000 plants that produce chemicals active against cancer cells. Seventy per cent of them come from the rainforest.

Rainforest plants are especially rich in so-called secondary metabolites, mainly alkaloids, which they probably produce to protect themselves from disease or attacks by insects. All such chemicals can have major and, on occasion, potentially useful effects on human physiology. Already researchers have turned up two plants – a Samoan tree and a Cameroon creeper – that appear able to fight HIV, the virus that causes AIDS. Rainforest animal products, such as the venom from snakes and spiders, can also have medicinal uses. One example is an effective anti-hypertensive drug derived from the venom of an Amazonian snake.

To a large extent, the search for new drugs from exotic plants and animals is a random process of trial and error – with an element of inspired guesswork. Yet insight and knowledge once existed. Many a pharmaceutical company from the developed world has had cause to regret the passing of local Third World cultures with their understanding of folk medicine. In Brazil alone, since the early 1900s, European colonists have already destroyed more than ninety indigenous tribes, each with a distinct culture. With them have gone centuries of accumulated knowledge of the medicinal value of forest species. Almost belatedly, the purveyors of Western drug-based medicine are courting the remaining witch doctors, shamans and tribal healers, trying to catalogue their knowledge of plants and herbal medicine before it is too late.

Many widely used drugs have originated from tropical species first used by forest dwellers. Quinine for malaria, curare as a muscle relaxant and vincristine for the treatment of leukaemia are well-known examples, but there are thousands of others – and probably many more to be discovered. In the Amazon, local inhabitants treat fungal infections of the skin with a particular sap and distinguish 250 types of dysentery, prescribing a different treatment for each type. In northern Nicaragua, clinical trials have confirmed that the local remedy, camomile, is effective against diarrhoea, a major killer in Third World countries. The list goes on, providing decades of research material for pharmaceutical companies – as long as the plants they require continue to be available.

But whose plants are they? This question has sparked another confrontation between Third and developed worlds. Not surprisingly, most of the technical expertise and commercial drive in the search for new drugs is based in developed countries. Such nations argue that drug-producing plants are a global heritage, which they should have the right to exploit; Third World countries, such as Brazil, Indonesia and India, within whose boundaries the plants live, consider that they are a local resource that only they have the right to exploit. They suspect that food and pharmaceutical companies based in the rich northern countries are angling to make most of the money from this 'global resource'. As one spokesman said, 'The high-sounding plea of the common heritage of humankind is a rhetorical device to disguise continued exploitation.' Whether their forests provide plants for food, drugs or genes, Third World governments consider that they have ownership and that any exploitation by outside interests should involve payment and 'royalties'.

Accepting this latter principle, some agencies have established themselves in the developed nations and, in effect, act as brokers for Third World countries, finding them buyers for their medicinal plants among the pharmaceutical companies of industrialized countries. In this way, plants from Ghana, Malaysia, Thailand and China have already found their way to Europe and North America. The suppliers and the brokers receive an initial payment for the plants, and then the companies screen the material for biological activity. If this leads to products, the suppliers are supposedly guaranteed royalties in return for their materials. But given that only one plant in 10,000 will lead to a commercial product, companies are often unwilling to fund speculative research and are always looking for ways to reduce their costs. The fear in the tropics is that within a few decades, biotechnologists will have ransacked their countries for genetic resources, taken them home and hoarded them in gene banks. After that, the original Third World resources will no longer be needed – or qualify for royalties.

One such story, which concerned a food rather than a medicine, began when it was found that a strain of cowpea being cultivated in northern Nigeria was unusually resistant to weevils. These voracious beetles often eat a large part of the cowpea harvest. The institute that made the discovery employed British researchers to isolate the source of the resistance, which turned out to be a molecule that interferes with the weevil's metabolism. A company at the university accommodating the research took out a patent on the gene that codes for the molecule, and began licensing seed companies to incorporate it in a number of different crops. Sales of genetically engineered weevil-resistant crops may still be a few years away – but Nigeria doubts it will ever receive the royalties it believes it is due. After doing their research on Nigerian plants and identifying the important gene, the British researchers managed to find the same gene in a Californian variety of cowpea. Thereafter, they used that instead.

Alternatives: Reforestation, Virtual Reality and 'Frozen Zoos'

Even if the science of global warming does eventually turn out to be solid, there are solutions other than the preservation of Third World forests. The developed world could undo its own misdeeds: a determined rather than a token reduction in carbon-dioxide production combined with northern reforestation would solve the mooted problem. Similarly, the argument that habitat destruction will make the world a boring place for future globe-trotters seems a flimsy reason to ask Third World people to starve when again there are alternatives. Films, books and zoos that even now provide the nearest thing to first-hand experience of wildlife and habitats for most people will continue to be available. Moreover, that experience could become increasingly vivid, thanks to the expected future of virtual-reality technology and ever more imaginative zoos and parks. It will also become more comfortable, safer and much cheaper than the real thing and will not involve robbing other people of the chance to use their environment to their own best advantage. Purists will complain, but there are lives and livelihoods at stake.

The final argument for conserving huge tracts of natural habitat – so as not to lose the potentially valuable genes carried in plants and animals – also looks increasingly weak. Again there is an alternative – to conserve only the genes, not the plants or animals themselves. This would allow a species' genetic material to survive even after the species and its habitat has become extinct. Any future company or researcher who comes to suspect medicinal or other uses for such genes would then have easier access than if those genes were running wild in a distant rainforest.

Genetic conservation not only offers the chance of warding off extinction without robbing local people of large tracts of land and resources, it also provides the means of resurrecting the species in the future if it ever seems a good idea. Part of the technology needed for the resurrection of extinct species is the cloning technology that created Dolly the sheep. No extinct or endangered species has yet been cloned through to adulthood. In 1998, though, scientists in New Zealand cloned the lone surviving member of a rare breed of cow. Like Dolly and the other animal clones to date, the cow's clone was created by transferring genetic material from an adult animal into an egg cell from a female of the same species. In the case of the cow, there was little problem finding either an egg donor or a surrogate mother to carry and give birth to the clone – there were plenty of cows to hand. It will be more difficult to find either for an endangered species – and even more difficult for an extinct species!

One of the first endangered species to begin the cloning process is the giant panda. A Chinese team has successfully produced a cloned panda embryo using cells taken from an adult animal. This is as far as the process has got, but it's a start. Panda eggs are hard to come by, so instead the Chinese team used the egg of a Japanese white rabbit to create their embryo from panda DNA. This made their job relatively easy, because the reproductive biology of white rabbits is much better understood than that of pandas. It remains to be seen whether using a surrogate egg from another species will greatly influence the animal produced. The search is now on to find a surrogate mother to carry the cloned embryo to term and hence produce the first cloned giant panda. Once again, the Chinese team believes that the best chances of success lie in finding a non-panda surrogate mother. They need one the right size and with similar enough reproductive traits: the black bear has been suggested.

The chances of success for the panda enterprise are currently slim, but cloning technology is improving all the time. The best strategy might be to put the panda cells into cold storage until such time as the technology is totally reliable. The same is true for all endangered species. The key to species and genetic conservation could be to put cells on ice now – in other words, to establish a 'frozen zoo' – and the process has already begun. In 1975, a programme began in the United States to preserve cells from endangered species. The San Diego zoo now has cells from over 4,000 individual animals, comprising over 400 species and subspecies within fourteen mammalian orders. They are stored in liquid nitrogen, which preserves DNA structure and genetic viability. Other frozen zoos are also springing up. When – if – cloning does become totally reliable, frozen zoos will be an invaluable source of genetic material for creating new animals. Exactly when the frozen genetic material will be used to produce animals in the future will depend on advances in resurrection technology but – straight from *Jurassic Park* – the first attempts are being made for animals already extinct.

The Australian Museum recently launched a foundation to focus resources on cloning a thylacine – also known as a Tasmanian tiger or wolf. Thylacines became extinct on mainland Australia around 1936 – the species received government protection two months after the last one had died. Dead specimens have since been stored in ethanol, a DNA preservative. The resurrection team is trying to create a genetic library of the thylacine's DNA, which would be stored in the genetic material of viruses or bacteria. If the chromosomes are badly fragmented, they will need painstaking reconstruction. If not, cloning could begin soon.

Currently, as planned even for the endangered giant panda, extinct animals can only be resurrected using a different species as a surrogate mother. This is possible. A bit of transplant surgery on embryos from both horse and zebra, for example, 'tricked' a horse surrogate mother into accepting a developing zebra embryo as if it were her own species. Before long, though, it may be possible to use a surrogate that is not another species – but a machine. So far, the only animals to have begun life in an artificial womb are goats. In 1992, Japanese scientists removed a goat foetus from its mother by Caesarean section after 120 days' gestation, about three-quarters of the way to its full term. It was then placed in a rubber womb filled with artificial amniotic fluid, and successfully 'delivered' seventeen days later. Eventually, such artificial wombs may make the perfect surrogate for the resurrection cloning of extinct species.

Frozen herbaria can be produced just like frozen zoos. Genetic material can be stored in gene banks in either their original cells or separately in bacteria or viruses. And as described for GM foods in the next chapter, it is now relatively straightforward to resurrect plants from cells and genes. Growing the plants will then simply be a case of giving them the right soil and atmosphere. They can be grown in greenhouses, nurseries or in parks, depending on their main use.

Resurrected extinct animals, though, are destined to live only in zoos or parks. With their natural habitats destroyed decades or so earlier, resurrected animals are unlikely to survive if released into the wild. This is the conclusion of recent captive-breeding programmes on endangered species that then attempted to release the animals back into native habitats. Such programmes have met with more failure than success: in one appraisal of 145 captive-breeding and reintroduction programmes, there were only sixteen cases where wild populations had been successfully established. Resurrected animals, following habitat destruction and extinction, are unlikely ever to become established again in truly unmanaged habitats. Nevertheless, they may always have a place in the zoos and parks of the future – and even if they die out again, there will always be further cells in frozen zoos to clone and resurrect. Aesthetically, this may seem a poor substitute for exotic animals that are born and run free, and it will offer little to the eco-tourist. But for many people, it may seem a better option than a full-blown mass extinction.

These futuristic forms of conservation may seem far-fetched, even macabre, but in the long term they are probably more realistic than the current programmes for habitat conservation – because the developed world seems unlikely to get its way. Rainforests are unlikely ever again to cover huge areas of the tropical land surface – any more than northern forests are ever again likely to cover temperate land surfaces. Eco-tourism is never realistically going to provide stable enough funds to sustain more than tiny pockets of untouched exotic habitats. And once pharmaceutical companies have stocked their tropical gene banks and bled shamans of their knowledge, they will no longer consider it cost-effective to sustain large areas of Third World forest. Barring an overnight miracle improvement in the wealth and health of Third World countries and a consequent immediate drop in birth rates, nothing conservation science can do – apart from the perfection of cloning and other biotechnologies – seems likely to halt the seventh extinction.

7

GM Foods:
Monstrous Saviours

In the 1980s, they were hailed as saviours. They would revolutionize agriculture, feed the world, save the environment, and give consumers food tailor-made to their needs. 'They' were genetically modified (GM) foods – and future generations would wonder how their antecedents managed without them. Then in Europe everything turned sour. In the 1990s, pressure groups showed their teeth. Word was spread that animal genes were to be inserted into plants, the environment would be destroyed, wayward genes would pepper food chains and human health would suffer.

While the United States looked on bemused, the European media had a wonderful time stirring up evangelical zeal. GM foods were renamed Frankenstein foods – and the crusade escalated. Consumers, alarmed that they had been eating monsters without even knowing it, demanded reassurances over safety and called for all GM food products to be labelled. Environmentalists demanded reassurances for the environment – but destroyed the crops that might provide reassurance. Organic farmers demanded protection against having their crops genetically contaminated. And a British Prince – who also happened to be an organic farmer – proclaimed that only his God has the right to tinker with genes.

Unlike that other environmental demon of the 1990s, global warming, which was remote, boring, and difficult to blame on anybody except everybody, Frankenstein foods had everything needed for lasting hype. They were 'here and now' and they were being forced on a hapless public by environmental ogres such as big business, scientists and governments. The media and pressure groups had the perfect vehicle to manipulate people's perspectives and emotions while ignoring the science.

So what are the facts of the matter? Do we need GM foods – do we want GM foods? How dangerous are they to human health and to the environment? Are they monsters as named, or saviours as promised? And – if the science is really as devilish as Europeans have come to believe – why is the United States so unconcerned?

Atlantic Divide

In many ways, the Atlantic divide – the difference in attitude between Americans and Europeans – is one of the most fascinating aspects of the GM saga. Although European countries import GM produce, only processed foods are allowed, not raw – and governments are under pressure to justify every tiny decision. European consumers are deeply suspicious, even afraid, of any food that contains GM materials. Hence the demand for reassurances and labelling and aggressive actions such as the destruction of trial GM crops and the blockading of ports that receive American produce. Many people would like to see a complete ban on imports and at least a five-year moratorium on growing GM crops for research purposes to give time for reassurances over safety to be given. No crops are yet grown commercially and, in a 1999 British poll, only 1 per cent of respondents expected to see any benefits from genetically modified food. So antagonistic – and unfocused – is the atmosphere that scientists asked to sit on advisory panels are subject to threatening hate mail, and firms trying to produce cheap vaccines for Third World countries are unable to raise the necessary money because their process involves GM plants. Such humanitarian projects have had to be abandoned.

In contrast, the average American has so far been blasé about consuming GM food and bemused by the European furore. In a policy statement issued in 1992, the American Food and Drug Administration (FDA) ruled that GM food did not need special testing or labelling. This must have reassured the American public because a 1999 poll showed that more than two-thirds of people would happily buy produce enhanced through biotechnology. They were so unconcerned that although 73 per cent had heard 'something' about biotechnology, only 40 per cent realized that GM food was actually on sale in supermarkets. By the turn of the century, up to 8 million hectares of maize – about a quarter of the country's entire crop – was genetically modified. Over forty GM plant varieties have now been approved for human consumption and these include raw foods such as tomatoes, papayas and squashes. Nor is the United States alone. In China, hundreds of millions of acres are being planted with GM wheat and soya for human consumption. The result is that very soon, more people in the world will be eating GM foods than not.

So why the divide? After all, the science behind GM foods is the same no matter where people live. Is it simply due to differences in national character? Perhaps Europeans really are more easily panicked than Americans – more distrusting of novelty, change, scientists, politicians and big businesses. Or maybe the European media gain more mileage from creating scares – or quasi-evangelical causes – than their American counterparts. The difference can't be due to the lack of pressure groups in the United States. Not only do they exist but they are also active, filing lawsuit after lawsuit in an attempt to slow down or reverse the GM snowball. Yet they have had little response so far from either the government or the public – not by comparison with their European counterparts.

So perhaps, despite the fears and the high principles, the difference is purely commercial: on the GM balance sheet, American companies stand to profit and European companies stand to lose. This commercial divide was graphically shown at the meeting of the World Trade Organization (WTO) in Seattle in December 1999. Outside the hall, the conflict was between riot police and protestors; inside it was between the Europe-led delegation and the 'Miami group' of nations, the United States, Canada, Australia, Argentina, Uruguay and Chile. The Europe-led delegation wanted the introduction of genetically modified food, plants and animals to be regulated by the UN's Biosafety Protocol, with environmental considerations paramount. Drafted after the Earth Summit in Rio de Janeiro in 1992, this protocol says that countries should be able to ban any GM material they consider to be a threat to the environment. But the Miami group didn't want trade in GM products strangled by environmental considerations, and so argued that the WTO should handle matters.

Much of the focus of the vitriolic debate around the world has been on whether foodstuffs containing GM produce should be labelled; the UN's Biosafety Protocol advocated universal labelling of GM foods. This is also the area in which anti-GM pressure groups have had greatest success as first Europe, then other countries acceded to the demands. Even the American Secretary of Agriculture conceded in 1999 that genetically modified food might one day have to be labelled – a move that some observers likened to the Pope reviewing his stance on contraception.

GM food manufacturers have always objected to labelling because it implies that their produce is in some way dangerous and so could have an adverse effect on sales. GM opponents, though, argue that consumers should have a choice over what they eat. We can return to this demand towards the end of this chapter – because the choice should be an informed choice. How many people – whether hostile Europeans or blasé Americans – can honestly say that their stance over the issue is founded on a clear understanding of the potential benefits and potential hazards?

Tasty Saviours . . .

The world is full of useful genes. The problem is, they're not always where they would be most useful. So why not move them around a bit and in an act of precision genetic engineering put them exactly where they are needed? This was the dream and GM food was the realization. The first research began to roll in the 1970s and the first transgenic – genetically modified – plants were created in 1983. The possibilities seemed endless and all seemed harmless and desirable. Innocent visions ranged from tomatoes that ripened without going soft and coffee plants that produced naturally decaffeinated beans to plants that produced their own insecticides and crops that could withstand global warming.

Of course, the idea seems weird: that a gene evolved to do a job in one plant or animal can actually do the same job if placed in another, but to modern geneticists, it's not at all strange. Genes may have been imbued with a magical element, but essentially each one is simply the set of instructions needed to manufacture a single protein. The gene itself is a length of chemical – deoxyribonucleic acid (DNA) – that is both simple and infinitely complex. It is simple because it consists of only four different types of chemical unit, called bases. It is complex because, when these are repeated over and over again in different orders and combinations they form a physical code – the genetic code. This code is variable enough to specify the order in which amino acids, the building blocks of proteins, should be joined together. A gene will make its particular protein no matter what species of plant or animal it is in, as long as it is 'switched on' and as long as enough of the necessary raw materials to make the protein are available.

Genetic engineering involves taking a gene from a cell in one species and inserting it into a cell in another species. When the gene is placed in a new cell, because it changes the proteins produced by that cell it can change the cell's function. It may change the way the cell works, or it may change the nature of any chemicals that the cell secretes. It is even possible to swap genes between plants and animals. Most plants and animals are very similar, genetically. Humans, for example, not only share 99 per cent of their genes with chimpanzees, which perhaps isn't so surprising, but also share about 90 per cent of their genes with an oak tree, which is. It's the genes they don't share that make all the difference, of course, but even these often do a large part of their job by calling up the services of genes that are shared. All that's needed then is to take a gene from one plant or animal, insert it into the chromosome of another and there's a very good chance it will carry on doing the job in its new host that it did in its old.

Tinkering with Nature

There are several ways of tinkering with a plant's or animal's gene plan – and most were in action long before the first genetic engineer appeared on the scene. The oldest tinkerer is evolution, which has been messing around with genes ever since the first self-replicating molecule appeared in the primordial soup. If it hadn't, we'd all still be in that soup. One of the life forms that evolution produced is nothing but a genetic engineer. Viruses make their living by changing other organisms' DNA. They invade a cell, insert their own DNA (or RNA) into the host nucleus and reprogramme the host into manufacturing copies of the virus instead.

Humans became dissatisfied a long time ago with the speed at which evolution worked and the opportunities to help humankind it had missed. We have been trying to improve on evolution's products for centuries – if not millennia – via organized breeding programmes. Much of this simply involves selectively breeding from the biggest and best, but some of it involves mixing genes from different species. Modern strains of wheat, for instance, have been genetically altered over thousands of generations by repeatedly crossing selected grasses together. As a result, wheat now contains a large selection of genes that originated in a species of wild rye grass. Even animals have had their genes mixed. Horse genes, for example, have traditionally been mixed with donkey genes to produce mules. Only the mule's sterility prevented it from becoming a traditionally produced Frankenstein's monster. In fact, modern humans probably eat nothing that has not been deliberately altered or that deserves the label 'natural'. There is little that is natural about a cow that can produce 10 gallons of milk per day.

Traditional methods, though, are also fairly slow; the results are unpredictable and often poor. The methods are also limited, being totally dependent on fertile matings of closely related species or varieties. This was particularly restrictive in modifying animals; plants are rather more liberal in their sexual activities with other species. Genetic engineering, though, overcomes this limitation and allows an individual gene to be deliberately chosen and inserted into an organism, greatly increasing the rate at which changes can be brought about. It also allows genes from one species to be introduced into another distantly related species, a process called transgenics.

Several methods have been devised to transfer genes from one species to another. The first transgenic plants were created by exploiting a natural genetic engineer – the bacterium *Agrobacterium tumefasciens*. The micro-organism normally lives in the soil, but it can infect plants that have suffered an injury. It injects part of its own DNA into cells around the wound and instructs them to manufacture further cells. The DNA also instructs them to make certain chemicals that the bacterium uses as a source of energy for its own growth and reproduction. Genetic engineers exploit this naturally evolved ability. Using modern biochemical techniques, they can put any gene they want into the bacterial DNA – a process much easier with a bacterium than a plant or animal. The bacterium then does the rest. Usually, the bacterium is tricked into injecting the required gene into plant cells in culture, rather than a growing plant, and once the gene has been successfully transferred, the cells are grown into a whole plant, which then is allowed to reproduce.

Other methods, collectively called 'direct gene uptake' methods, involve simply adding the gene to cells in the hope that it gets incorporated into the host's DNA when cells divide. One approach is the 'chemical' method, which uses magnesium and calcium ions and polyethylene glycol to make the membrane permeable to DNA. Another is to use electrical pulses to create pores in the cell membranes. Microinsertion of DNA directly into a cell is a third approach. This can use either simple micro-injection or microprojectiles. The projectiles, small particles of tungsten or gold coated with DNA, are accelerated by a particle gun or electric discharge so that they penetrate the outer cell wall and membrane of intact cells. Injecting DNA into developing pollen cells has produced transgenic rye plants. Shooting micro-projectiles into soya bean embryos has produced transgenic soya bean plants. These methods deliver DNA directly into cells of the plant's sexual-reproduction system and allow transformed plants to be grown from seed without the need for sophisticated tissue cultures.

Saving People

The first GM food to go on sale in the United States was an indulgence rather than a matter of life and death: a transgenic tomato plant, the fruits of which soften more slowly than usual. An enzyme that contributes to the softening of the fruit during ripening was suppressed by introducing an engineered gene. Tomatoes are often picked green and ripened artificially using ethylene. If softening is delayed, growers can have tomatoes that they can allow to ripen naturally, perhaps improving flavour, without worrying that the fruit would turn into goo before it was sold. Other developments, although still not a matter of life and death, have rather more global impact than that achieved by altering the taste of a tomato. For example, Japanese researchers have managed to slash levels of the major allergenic protein in rice by 70 to 80 per cent by inserting a gene that blocks the protein's production in the plant.

One of the early dreams for genetic modification was to supply poor countries with cheap vaccines against diseases such as hepatitis B or cholera in a form that didn't need expensive refrigeration to store or the inconvenience of needles to inject. Using plant viruses engineered to produce some of the protein fragments from disease-causing organisms, it has been possible to create food plants, such as potatoes and bananas, which stimulate the immune system when they are eaten. Clinical trials have been carried out on a potato-based vaccine against hepatitis B, a virus that triggers liver cancer and causes 1 million deaths worldwide each year. Similar vaccines are being developed against cholera and other diarrhoeal diseases. The ultimate aim is to develop bananas containing the vaccines, which could be grown cheaply and then eaten to stave off killer diseases in the world's poorest countries.

Among the first diseases to be attacked in this way was a strain of the bacterium *Escherichia coli* (better known as simply *E. coli*), which causes food poisoning in humans via a protein that binds to its victim's gut cells. The gene that produces the protein was transferred to *Agrobacterium tumefasciens*. This in its turn was used to insert the gene into a potato plant. Once the foreign DNA segment was incorporated into the potato's own DNA, the bacteria were killed off with antibiotics, leaving just a potato plant that produced the same protein as *E. coli*. When the potatoes were fed to mice, the rodents began producing antibodies to the *E. coli* poison, but suffered no ill effects from digesting the poison protein itself, which was probably broken down in the mice's stomachs. Tests have since begun on human volunteers. The only problem with growing potatoes to produce vaccines is that cooking tends to destroy the protein component of the vaccine, so they must be eaten raw. Bananas would be a better option, one banana potentially producing a whole range of different vaccines.

Another way in which genetic engineering could save people is by increasing the range of foods that could be grown in Third World countries prone to famine. A simple redistribution of what we grow now, even if it were possible, would not feed the 10 billion humans expected by 2050. Something needs to be done – because traditional methods of improving crops seem to have gone about as far as they can. After 5,000 years of selective breeding, further major increases in yield seem unlikely.

Most of the world's food calories come from grain. The biggest opportunity for increasing grain yields is to produce varieties more precisely adapted to local conditions. Genetic manipulation could make plants grow better in a wider range of conditions. Crops could be made more efficient at photosynthesis, could increase their yield, and decrease their dependence on fertilizers by enabling them to take nitrogen from the atmosphere. Currently, at least 40 per cent of crops in marginal habitats can be expected to be lost to weeds, disease and insects. Genes can be introduced for resistance to fungi, bacteria and viruses. Plants can be made more resistant to harmful radiation, extremes of temperature or pH, drought, flooding, heavy metals and various pollutants of air, water or soil. Crops that thrive despite drought and salty soils would let farmers expand production into marginal lands. And the nutritional content of staples could be improved. If maize, for example, can be made to produce more of the amino acids it naturally lacks, the 80 million people who live almost exclusively on maize would get more protein: it could almost be made into a complete balanced meal.

Few crops that benefit the Third World have actually been produced so far, though Vietnamese biotechnologists have recently used genetic material supplied by the International Potato Centre in Peru to develop potato varieties that can thrive in Vietnam's climate. They are now also at work on improved coffee varieties. In Thailand, the National Centre for Genetic Engineering and Biotechnology has developed a local starch industry using native cassava. Elsewhere, modified fodder crops such as corn or soya are in the pipeline. These contain more calories, so that when fed to livestock more meat can be produced for each hectare grown of the crop. Also in the pipeline are crops that destroy toxins produced by moulds, such as fumonisin, which cause massive crop losses after harvest. Finally, disease-resistant crops, such as sweet potatoes and cassava, staples of the poor, are being produced which fend off viruses.

Saving the Environment

Traditional agrochemical methods are crude. Take an insecticide, spray it over a crop, and kill every insect that comes near. Or spray herbicide and kill every weed in sight for years to come. Genetic engineering can do better than that. It can produce crops that manufacture their own insecticide; it can produce crops that are resistant to herbicides that quickly break down in the soil.

The advantage of a crop that produces its own insecticide is that only insects that attack the plants are affected, leaving all the other inhabitants free to carry on their lives. Financial incentives are great: pesticides applied to control herbivorous insects cost some $3 billion a year worldwide, including $400 million to control the larvae of moths and butterflies in the United States alone. Protecting GM crops should be a lot cheaper: farmers should use less chemical spray, thereby also making the environment cleaner and healthier. The first transgenic crops to produce their own insecticide – tobacco and tomato – used the toxic proteins produced by strains of the bacterium *Bacillus thuringiensis*. They are referred to as Bt crops. These proteins protected them from damage caused by the larvae of a moth, the tobacco budworm. Another genetic modification made tobacco plants more resistant to this caterpillar by giving the insect indigestion. A gene was introduced that coded for an inhibitor of trypsin, an enzyme that breaks down proteins during digestion. Such 'protease inhibitors' may be able to help plants fight off a much wider range of insects than do the toxins from *B. thuringiensis*.

Crops resistant to specific environmentally friendly herbicides could also be beneficial. Farmers now use more than 100 herbicides as weedkillers, and there is increasing concern about their persistence in the environment and lasting toxicity to animals, including humans. Potent herbicides – such as glyphosate and phosphinothricin – do exist that are not very toxic to animals and that do readily break down in the soil. Unfortunately, they kill a wide range of crop plants as well as weeds. Transgenic crops resistant to such herbicides, though, would allow farmers to do less lasting damage to both weeds and animals. Most herbicides kill plants by inhibiting either their ability to synthesize essential proteins or by blocking photosynthesis. Genetic engineers have so far produced herbicide-resistant plants by introducing DNA that makes them overproduce the proteins that are sensitive to the herbicide (so absorbing the herbicide's impact), or makes them produce proteins that degrade or detoxify the herbicide.

... Or Toxic Monsters?

Not knowing the real potential hazards from GM foods, many people have been frightened into fantasies straight from science fiction. Phobia is rife and has been fully exploited by the media. So much so that in the average European's mind GM plants have taken on triffid-like stature. For example, in 1999 a British newspaper criticized scientists at the nation's biggest biotech company for eating GM tomatoes as part of a stunt to show they were safe. The main worry on the newspaper's mind was that seeds from the tomatoes might pass straight through the scientists and germinate in a sewage farm, thereby threatening the population at large. The implication was that with just one such modified monster out of jail, the whole of humanity was in jeopardy.

Health Hazards

Frankenstein foods have triggered many fears, some of them rational and some of them not. The irrational ones owe more to such media hype and the lasting power of Mary Shelley's imagination than they do to science. Suggest to some Europeans that scientists have put a fish gene into tomatoes and the chances are that they will worry about growing fins. The fact that the same gene has always been in their diet – in fish – would do little to ease their phobic vision of it in a tomato. A moment's reflection, though, should reassure everybody that the genes of the plants and animals we eat do not normally hijack our genome and turn us into fish or tomatoes. Such food not only lacks the genetic apparatus to force one of its genes into our genome but our digestive system also has – of course – evolved to prevent it happening. Even viruses and bacteria, the most likely organisms to have the apparatus to attack our genes, cannot manage it without millennia of evolution behind them to help – long enough to produce 'coats' specifically designed to resist the digestive process. Studies of mice show that digestion shatters chromosomes from food into pieces too small to contain intact genes – even when those chromosomes originate in viruses. Some snippets of viral genes do get into the animal's bloodstream and cells – and, on rare occasions, they may even link to mouse DNA – but no evidence has yet been found of active ingested genes. This includes genes actually designed to work in human cells.

Super Microbes

The more rational fears over human health have not revolved around the trait-gene being inserted but around the genetic construct that has been inserted around it. The production of transgenic plants is a hit-and-miss affair and the first thing genetic engineers have to do is identify which of their targets – cells, seeds or plants – has successfully accepted the trait-gene. The method used is to install a marker gene alongside the trait-genes – and originally these markers were genes that conferred resistance to a particular antibiotic. The cells or seedlings could then be exposed to that antibiotic. Those that have successfully accepted the gene bundle are resistant and survive; those that have not, die. Once the plants have been screened, the resistance gene serves no further purpose but cannot be removed. The fear is that when people eat GM foods containing antibiotic marker genes, some of the bacteria in their guts could pick up these genes during digestion and become resistant to the antibiotic concerned. If the bacteria are potentially harmful (most bacteria in the human gut are harmless or even beneficial), the result could be an epidemic of gut diseases that might be difficult to control with antibiotics.

GM technology has always assumed that microbes in the human gut are as unlikely to pick up genes from food as humans themselves – otherwise the antibiotic markers would not have been developed in the first place. As we have just seen, most DNA from food is destroyed well before it reaches gut bacteria, with any surviving remnants being shredded again inside the bacteria by so-called restriction enzymes. Even if intact genes were successfully to invade a bacterium, they would be unlikely to spring into action because DNA switches designed to work only in plants will control their activity. On top of this, either the antibiotic marker genes used in crops approved for commercial growth relate to antibiotics of no clinical importance or the genes have been scrambled, so there is little chance for their resurrection in another host. So it was always thought unlikely that these particular genes could boost the spread of antibiotic resistance in human pathogens.

Even so, few genetic engineers would dare to say it is impossible for resistance genes to jump from GM foods to bacteria in the gut – and the possibility has recently been supported by research in the Netherlands. Using an 'artificial gut', Dutch researchers have shown that DNA may remain intact in the large intestine for several minutes longer than originally thought: DNA from bacteria survived for about twelve minutes in the large intestine. This produced a one in 10 million chance of DNA containing antibiotic-resistance genes being passed to any single gut bacterium. Such odds may sound low – but there are normally around 1,000 billion such gut bacteria, suggesting that many might be transformed. However, DNA from a raw GM food – a tomato – did not transfer to indigenous gut bacteria at a detectable level, even though up to 10 per cent of the food's DNA reached the colon.

Digestion, though, is not the only way of deactivating genes: cooking does the same. It was partly for this reason that Britain and some other European countries allowed GM foods for human consumption only if they had already been processed – tomato paste, for example. Raw foods have so far not been allowed. This is despite the evidence from one huge – albeit uncontrolled – experiment that seems to show that even raw foods are safe. Americans have been eating raw GM foods since 1983 with no reports, so far, of any epidemics from newly resistant gut microbes.

The risk of bacterial resistance, though, should in any case soon fade into GM history. Antibiotic markers are not essential to the GM process and already, since the European furore began, two alternative methods of marking genes have been devised. One, developed initially for lettuces and tobacco, is to replace the resistance marker gene with a different type of marker gene that simply causes the plant to produce many new shoots when exposed to a particular steroid. The other – successfully tested on maize, wheat, rice, sugar beet, oilseed rape, cotton and sunflowers – uses a sugar-based alternative. The new marker gene enables plants to digest a simple sugar called mannose-6-phosphate. Most plants can't handle this sugar, and so die when fed mannose alone. The marker gene should be safe – it already exists in many familiar crops, in most of the animals that humans eat, and in all mammals, including us. It also already exists in many species of gut bacteria. The only problem is that the technique cannot be used for all plants because some, such as soya and other legumes, already have the gene.

Poisonous Plants

Each promising transgenic crop is routinely put through a battery of biochemical checks to monitor levels of nutrients, proteins and potential poisons. In some cases, the crop is also fed to livestock to check that the animals gain weight at the normal rate and remain generally healthy. The GM food industry felt that it was doing everything necessary to ensure the safety of public health. In late 1998, though, any complacency was shaken by the announcement on British television that transgenic potatoes had been shown to be poisonous to rats. The scientific research on which the announcement was based was not formally published until much later in the year, but by then the whiplash had been felt throughout Europe.

The potatoes concerned had been genetically engineered to produce GNA, a type of protein called a lectin. GNA is made by snowdrops to deter sap-sucking insects – hence the potatoes should have had heightened resistance to insect attack. As part of a safety check, some rats were fed the modified potatoes, some were fed normal potatoes laced with lectin, and a control group was fed normal potatoes. Some rats ate the three types of potato raw, some boiled. After ten days, the rats were killed and examined. A variety of differences were found that indicated damage to the intestinal linings of those fed GM potatoes. The statistics suggested that the GNA gene itself was not to blame, so the authors pointed the finger at the genetic construct inserted along with the trait-gene. This construct included a genetic switch – taken from the cauliflower mosaic virus – designed to promote (turn on) the GNA gene. The alarm that ensued had less to do with this particular modification of the potato than with the fact that constructs containing this promoter are routinely used to genetically engineer plants. Perhaps, then, there was a whole range of GM foods on the market or in the pipeline that were destined to ruin the intestinal lining of any mammal that ate them.

In fact – as many people were quick to point out – the authors were not justified in drawing their conclusion. They had omitted to do the crucial comparison, which was to produce potatoes with only the construct and not the GNA gene and see if that alone was enough to cause the damage. Even had they done so, it's unlikely they could have proved their case beyond reasonable doubt because there were many other problems with the experiment. Changes in the rats' organ weights and immune reactivity showed no unambiguous association with genetic modification: starvation or known toxins in raw potato were the most likely causes of any changes seen in the rats. The rats became malnourished no matter which potato-type they were eating. Potato is not rats' favourite food and those in the experiment couldn't be made to eat enough to stay healthy. Alternatively, toxins – called glycoalkaloids – might have caused the gut changes.

There is no scientific doubt: this experiment could never support the claims made for it by its authors – nor those made on its behalf by those who dearly wished it to be true. Nevertheless, the immediate and almost universal condemnation from other scientists that followed the initial announcements actually backfired. The media easily made the response look as though the scientific establishment was ganging up to silence a messenger it didn't want the public to hear. The emotions raised by the work quickly transcended the actual science. There is no point now, no matter how justified, in criticizing the science or the researchers any further. The only useful way forward is to have the experiment repeated – but properly – to resolve whether there really is any substance to the fears aroused. The chances are that there is not – but it wouldn't be the first time in the history of science that conclusions from a badly done pilot experiment have later been vindicated.

New Allergies

The danger that GM foods might create new allergies is more difficult to counter than the danger of toxicity. When genetic engineers shuttle new genes into plants, they could inadvertently introduce proteins capable of triggering respiratory or inflammatory problems in the 1 to 2 per cent of people who suffer from food allergies. We know this because a new allergy was also one of the dangers of traditional selective breeding. For example, in the mid-1980s, plant breeders in the United States introduced what they thought was a wonderful new strain of celery. Highly resistant to insects, it promised to boost yields dramatically. There was just one small problem. People who handled the celery sticks began complaining of severe skin rashes. Dermatologists discovered that the celery was shedding psoralens, natural chemicals that become irritants and mutagens when exposed to sunlight.

In fact, the danger of creating allergenic foods is probably greater with conventional breeding than with genetic modification. Any one of the host of new genes introduced when, say, wheat is back-crossed with wild relatives could cause problems. Such traditional practices control neither the number nor type of the genes transferred. It is very much a process of trial and error. At least genetic engineers know how many genes they are introducing and what they are supposed to do. Moreover, because of the intensive testing that regulators demand for high-tech food crops, by the time a GM food reaches people's plates it is not merely as safe as a conventional food – in some respects it is actually safer. This doesn't mean, though, that there is no danger.

For example, in the early 1990s researchers engineered a more nutritious strain of soya bean by adding a gene taken from brazil nuts. The gene encoded a protein rich in a nutrient that is in short supply in ordinary soya beans. Research based on animal experiments published only a few years earlier had suggested that the same protein was not an allergen. Nevertheless, as a precautionary measure, a study was made of the soya bean's influence on the antibodies and immune responses of patients allergic to brazil nuts. The conclusion was that the transgenic soya was likely to trigger a major attack in such people, so the project was dropped. To critics of GM food it seemed like a narrowly averted disaster; to supporters it seemed to show that the validation process was working.

Environmental Armageddon

Whereas most concerned people are interested in the potential health hazards of GM foods, the most vocal opponents are environmental pressure groups. For the moment, all of the foreseeable problems are fears, not facts – and the biggest fear of all is that by the time any environmental dangers that might be posed by GM foods have been realized, it will be too late to do anything about them. Plants may not move very fast, but they are as capable of escape as any animal – and they are much more likely to be promiscuous with their genes. Potential problems are more likely to remain local than to become global, but apocalypse is not impossible.

Consider the tale of the Oxford ragwort in Britain, a small plant with pretty yellow flowers. The odyssey began in the seventeenth century with the introduction from Italy of a species of ragwort to the Oxford Botanic Garden. In the 1790s, the ragwort escaped and began to spread. On the way, its genes mingled with others; so much so that it spawned a new species and altered the genetics of a native flower. It has now reached north of the River Tay in Scotland – and is still on the move. Some people blamed the British outbreak of BSE on the changes brought about on British pastures by this escaped ragwort and its progeny. If this were true – which it probably isn't – and if half the human British population eventually die of nvCJD – which it probably won't – the blame could be traced to a bygone botanist and his tampering with nature. And if BSE has really reached the United States and southern hemisphere . . .

Many fear that the GM movement is seeding a thousand such disasters, though few would suggest that GM foods could cause human suffering on quite such a scale. Nevertheless, many are convinced that at least the environment is bound to suffer. The main perpetrators are expected to be the crops engineered either to be resistant to herbicide or to produce their own insecticides.

To Spray or Not to Spray . . .

Crops – such as soya and potato – engineered to resist specific herbicides will allow farmers to spray their fields with powerful chemicals that will kill everything botanic except the crop. From this simple development, environmental fears go in all directions. Naturalists fear that the process will encourage the expansion of the ecologically damaging monocultures characteristic of intensive farming. Moreover, as herbicides do not affect the engineered plants, farmers might be less careful in their use of these toxic substances. Some of the weeds likely to be attacked are not only threatened plants themselves, but they also provide important food or shelter for many insects, other invertebrates, birds and mammals.

In contrast to the naturalist's concerns, the farmer's greatest fear is the production of superweeds. Agriculturalists fear that herbicide-resistant genes from GM foods might be passed on to weeds, enabling them to grow out of control all over the country-side en route to spreading from farm to farm. There is also a subtly related fear – that big businesses will strangle farming choice. Take the Roundup Ready soya bean produced by a major American company. Genes have been inserted into the plants that make them resistant to the powerful herbicide Roundup, pro-duced by the same company. Farmers who use this combination of seed and herbicide have an easy way of ensuring weed-free crops – and the company collects on both seed and herbicide. Such combinations might solve immediate problems of cultiva-tion, but may also lead to farming dependence on relatively few combinations of herbicide-resistant crops plus herbicides. At this point, the fears of farmers and environmentalists merge. If the resistance spreads to weeds, we would have to develop new combinations of seed plus herbicide. Farmers would be locked into an arms race with the natural world and dependent on a few massive, high-tech agricultural companies to save them.

There are further dangers with plants that produce their own insecticide. In principle, these plants should be more environmen-tally friendly than traditional plants that can only be protected by spraying with chemical insecticides. The insecticide-producing plants should kill only the insects that attack the plant, not all those that happen to be in the field. Opponents, though, point out that not only the pests but also their predators will suffer – including more widely beneficial predators such as ladybirds. They also argue that the long-term effects of GM crops could be quite different from those of chemical sprays because the latter are used intermittently, allowing insect populations to recover.

Without solid evidence either way, the arguments for and against GM technology always seemed finely balanced – until they were tilted by the claim that one of the victims of engineered crops in the United States might be that doyen among insects, the monarch butterfly. The population of monarch butterflies that lives east of the Rocky Mountains overwinters in Mexico. It flies north again in spring to breed in the United States and Canada, particularly in the 'corn belt' of the Midwest, where GM maize is grown. Monarch caterpillars feed on milkweed plants, and when these grow near maize their leaves become covered with maize pollen. When pollen from maize engineered to produce Bt toxin was sprinkled onto milkweed leaves in the laboratory, almost half of the monarch caterpillars fed the leaves died within four days. None of a similar group of caterpillars that ate leaves sprinkled with pollen from unmodified maize died.

The company that produces the most widely planted variety of Bt maize in the United States knew that butterflies could be affected by Bt toxin but did not expect the danger to be great in the field. The question now is how much pollen is actually spread in the wild, how long it stays on milkweed leaves and whether it is produced when caterpillars are feeding. The company is embarking on such field studies – but critics say that this should have been done before farmers in the United States planted millions of hectares of Bt maize. Environmental activists and organic farmers have already filed a lawsuit against the Environmental Protection Agency, saying it approved Bt crops without sufficient review of environmental safety.

Pollen Fever

The monarch saga – like the ragwort saga centuries earlier – revolves around an unlikely villain – pollen. So, too, do other concerns about herbicide- and insecticide-resistant plants. First, there is the environmental concern that wild plants closely related to the modified crop will be fertilized by modified pollen. This could produce wild hybrids that are resistant to herbicides – superweeds that will overrun the countryside, outcompeting natural plants. Secondly, organic farmers fear that pollen from modified crops will contaminate their crops, making their produce unsaleable under the rules of organic farming. Whereas currently some contamination with pesticides does not threaten organic status, contamination with a few GM pollen grains does.

The nature of pollination, combined with a certain readiness of plants to hybridize, has always endowed the potential for unusual sexual unions. On the farm, hybridization is most likely when a flowering crop grows in proximity to wild relatives. These may be either native plants, or weeds that escaped from the crop in previous years and have taken up an independent existence. Such hybridizations are not new – the advent of genetic engineering has merely brought them into the public eye. Plant biologists believe that genetic leakage has occurred in the past in a wide range of crops, such as sorghum, oilseed rape, apples and sugar beet. Yet there is a shortage of well-documented cases. When the Royal Commission on Environmental Pollution examined the matter recently in Britain, it found no evidence that traits had spread from traditional crops to wild relatives. However, little real research has been carried out – and until such research gets under way, ideas about the frequency of genetic leakage from crops and its impact on wild populations will remain unsubstantiated.

Particularly problematic is the aerial transport of pollen grains. With plants that are pollinated by bees, the distance travelled by pollen in any one year will usually be small. Honeybees and bumblebees are often loyal to a particular variety of flower at any one time. Moreover, they prefer short trips between blooms, visiting several flowers on a single plant, perhaps, and then flying to a near neighbour a few metres away at most. Other insects, though, are a different matter. Butterflies, for example, travel up to twelve times as far as bees between successive stops within a field – and some butterflies, like the monarch, migrate hundreds or even thousands of kilometres in their lifetime. Long-distance pollination is also possible when pollen is transported by the wind. Again, most pollen is deposited near to home, but a small amount can travel for miles – hence urban hay fever. A pollen trap taken up in an aircraft and flown out over the North Sea between England and the European mainland picked up bursts of pollen from England up to 300 miles out to sea.

Clearly, pollen grains can travel huge distances. Most of the long-distance travellers, though, will probably never contact a receptive flower, so from a genetic point of view they are impotent. Quantifying this expectation is difficult: it is easier to find the proverbial needle in a haystack than to find the successful long-distance pollen grain, given something so small spreading over an area so large. The best that research has managed so far is to demonstrate gene flow over only modest distances. Using various markers, wind-pollinated plants such as grasses have been found that must have been pollinated over a distance of 150 metres. Pollination over much longer distances, though, must happen regularly. Traditional plant breeders have known for decades that if a plant variety has to be kept genetically pure, it has to be grown at a reasonable distance from any nearby relatives. For sugar beet or members of the cabbage family, the distance involved has to be at least a kilometre. It will be no surprise to find eventually that pollination, whether by insects or wind, occurs over distances of tens or even hundreds of miles.

Leakage of genes from crops into the wild has probably been going on since people first began to till the soil. Genetically engineered plants do not present a new problem, merely a variant on an old one. Hence, many researchers believe that whether a crop is conventionally bred or genetically engineered has little influence on the risk it poses to the environment. Rather, the risks depend on the genes the plant happens to possess, not how they got there. Critics of genetic modification disagree, arguing that engineered crops can carry genes from stunningly varied sources – unrelated plants, bacteria, viruses or even animals. However benign such genes might appear in the laboratory, leakage into the wild could present them with all manner of new evolutionary opportunities.

There is, though, a development in the GM industry that could solve this problem: the production of plants that are pollen free.

In the United States, GM green-hearted chicory and radicchio rosso seed that produce pollen-free plants have already been supplied to farmers.

Meaningless Labels: What's Really at Stake?

Clearly, organisms have been genetically modified for a multitude of reasons. Some are intended to help the consumer by improving taste or by making food healthier to eat. Some are to aid farmers by helping their fight against weeds and pests. Some are to help the environment by reducing the need for chemical sprays. Some are to help agriculture in the Third World. And some are purely for medical purposes such as the economical production of vaccines. Against this range of uses has to be set the range of potential dangers: primarily to human health and to the environment.

Given the many facets of the GM scenario, the global pressure for food to be labelled as simply containing or not containing GM food is revealing. The argument is that consumers should know what they are being offered so that they can make a real choice. But the choice should be informed, not Pavlovian. Not everybody worries about health risks from food. Even at the height of the BSE crisis in Britain, people ate beef and complained when the government restricted their choice. Nor does everybody worry about the environment. As long as food is safe to eat, some people care little whether insecticides or herbicides are within the food, as from genetic modification, or on top, as from sprays. On the other hand, people who eat organic produce may well object to one GM food that contains pesticide but not to another that has simply had one gene (protein) changed so that the plant ripens more slowly. Not a single food we eat is genetically the same as its wild ancestor and many organic consumers accept this. In contrast, others would not consider a GM food under any circumstance.

To make an informed choice, therefore, consumers need to be

told what genes have been changed, why they were changed, and what construct was used. They need to be told how much of the change will be broken down by processing, cooking or digestion. In addition, they need to be told in what way the plant was a proven – or potential – threat to the environment, and in which country. Then, depending on their personal principles, they could make an informed choice. The label would need to be large – and the consumer would need to understand the information. The result, though, would be the opportunity to make the educated choice that pressure groups say is what the consumer wants. But this isn't what has happened. It speaks volumes for the nature of the labelling argument that the demand is simply to have food marked as 'containing GM food' or not. The implication is that all GM food carries the same risks – both for human health and for the environment. This is blatantly untrue, which rather suggests that the demands for labelling are founded less on an understanding of the science behind GM foods than on a question of the principle: any tinkering with genes must be bad.

Testing Times

The potential benefits of GM foods are clear; so are the potential risks. The rational way forward is now to test both. There is much for scientists to do and the process isn't easy – because no matter what tests are carried out in the laboratory and no matter what reassurances scientists and governments might give, the final evidence won't come until about twenty years or so after GM foods have become a fact of life.

Safe Food

Consumers want reassurance that GM foods are safe. They don't want to turn into a tomato, they don't want to be poisoned, they don't want allergies and they don't want to become resistant to antibiotics. Apart from the tomato-phobia, this is a tall order –

because such reassurances cannot be given for any food, even those conventionally produced. Kidney beans, for example, are toxic; so are parts of crab. Dozens of people die each year from the cyanide in peach seeds. Onions – and grain – become toxic if particular moulds or bacteria attack them. Some people die from eating peanuts. And all over the world, people are becoming resistant to antibiotics – not because of what they eat but because the 'magic bullets' have been over-sought and over-prescribed. None of this has anything to do with GM foods. In fact, some GM foods could be engineered to rid us of these dangers.

The best that scientists can aspire to over the new produce is to offer reassurance that any new GM food is no more dangerous than the non-GM versions that went before. That isn't easy either. Of course, if something is really toxic there is no problem: a little is fed to a rat or mouse and the unfortunate animal dies. But if something is only toxic in large amounts such a test is almost impossible. When GM tomatoes were tested, for example, they were freeze-dried and fed to rats. But to obey existing toxicological guidelines, it was necessary to feed the animals so much that each rat got the equivalent of thirteen fresh tomatoes a day. Any more, and the rats would have been poisoned by the tomatoes' basic nutrients, such as potassium. Yet still the test hadn't fed enough to reach the toxicological criterion of 'safe'. Until a food is fed to people – and nobody dies – who can be certain that the food is safe? And if the danger is cancer, then it may take decades before the link is established. Even something as carcinogenic as tobacco was smoked and chewed for centuries before its dangers were widely accepted – and even now some people still refuse to believe that the case has been proven.

Potential allergies are even more difficult to screen: in part because people differ in their allergic responses; in part because what is allergenic to humans isn't necessarily allergenic to lab animals; and in part because considerable exposure may be needed before the allergic response is shown. Currently, the rule of thumb is that if a transgenic protein comes from a known allergenic food, it is subjected to immunological tests. If the protein comes from other sources, researchers study its molecular structure, looking for similarities with allergy-triggering proteins in the databases. The protein's chemical hardiness is also scrutinized. In test-tube simulations of the heat, acid and enzymes found in the stomach, most proteins are torn to shreds in seconds. Allergens tend to survive several minutes before they, too, are destroyed. It all sounds reassuring, but in reality allergenic science is in its infancy. There is still a danger that any new food – whether produced by conventional methods or by genetic engineering – will escape the safety net.

Whenever official approval for the introduction of (GM) foods has been given in Europe or the United States, regulatory committees have invoked the concept of 'substantial equivalence'. This means that if a GM food can be characterized as substantially equivalent to its 'natural' antecedent, it can be assumed to pose no new health risks and hence to be acceptable for commercial use. This seems reasonable in principle, but in practice is little better than asking a collection of scientists for their opinion. The concept of substantial equivalence has never been properly defined. How similar is similar? A new food clearly isn't identical to its predecessor otherwise it wouldn't be new – so how great a biochemical difference need there be for it to be different? For example, GM herbicide-tolerant soya beans have been deemed equivalent to their predecessors. Their chemical composition is different, of course, otherwise they would not withstand the application of herbicide – and in the laboratory it is quite straightforward to distinguish the particular biochemical characteristics that make them different. However, the genetic and biochemical differences discovered have been assumed to be toxicologically insignificant – and similarities in protein, carbohydrate, vitamins, minerals, amino acids, fatty acids, fibre, ash and other variables seem great enough for the new variety to be deemed substantially equivalent. This will reassure some, but not others.

Safe Environment

Reassurances over the environment will be even more difficult to provide than reassurances over safety; they may even be impossible. It wasn't possible to protect the environment from traditional farming, so it won't be possible to protect it from GM farming. Across Europe, farmland wildlife – especially birds – has been in decline for decades; long before GM crops caught the eye of pressure groups. The problem has been caused by the everyday use of highly toxic chemical sprays, by ripping up hedgerows, and by the relentless spread of intensive farming practices that we have never quite had the collective interest or energy to resist.

The hope is that GM farming will be better for the environment than the old; the fear is that it will be worse. Scientists must try to compare the two methods for environmental impact – and the first tests have already started. In Britain, the process of pitting biotechnology against conventional farming in a battle to see which is kinder to nature began in 1999. If the former fails to establish its conservation credentials, much of Europe may be tempted to consign genetically modified crops to the history books.

The British trials are focusing on the three GM crops – spring oilseed rape, winter oilseed rape and forage maize – that are closest to the marketplace in Britain: all are genetically engineered to withstand 'broad-spectrum' herbicides that kill almost any weeds. No herbicides of this strength have been used in Britain before. During the four-year trial, genetically modified crops are to be planted alongside conventional crops at various sites around Britain, either in each half of the same field, or in identical neighbouring fields. Armies of field biologists from government-funded research institutes will monitor weeds and wildlife in and around all the fields to evaluate any differences between the farming methods.

It has to be said, though, that such tests smack of cosmetic politics rather than robust science. Such small-scale studies will tell us nothing about the safety or danger of full-scale GM crop cultivation over large tracts of countryside.

Best Evidence

Nevertheless, the dangers or otherwise of GM technology will soon be known – because thanks to the Atlantic (and other) divides the populations and environments of the United States and China are already guinea pigs in the critical test of nation-wide exposure. These two countries' cavalier adoption of genetic biotechnology will provide the best possible evidence on the safety of the process. As yet, there are no signs of Americans or Chinese dying in the streets, of an increase in allergies, or of epidemics of antibiotic-resistant bacteria – but it is probably still too soon to fully relax. It is less than a decade since a range of GM foodstuffs came on the market in the United States and were grown in reasonable quantities – so it will probably be another twenty years before we know whether the farming revolution was a success, a disaster or a non-event.

In many ways it is surprising that pressure groups are still trying to change the American attitude to GM crops. According to pressure groups' own propaganda, there is no further point. In Europe, the crusade against small-scale tests involving only tens of fields claims that the moment such tests begin, harm to the environment may be irreversible. The genes will be out and there will be no going back. If this is true, then whatever anti-GM pressure groups achieve in the United States in the future it must be too late. The steamroller of biotechnology that is the American and Chinese experience may as well continue, giving the rest of the world the reassurance it wishes – or the evidence of harm it dreads. Yet the anti-GM lobby continues to step up pressure in the United States, not only over meaningless labels but also over the whole GM enterprise. The global impression is that opponents are not interested in reassurance, only in squashing the GM movement come what may; once again the driving force seems to be a generic dislike of the principle rather than a noble fear of the dangers. The motives behind such generic disapproval are obscure and maybe the opponents themselves do not always know their origin. Undoubtedly, some of the opposition stems from religion and the feeling that humans shouldn't meddle with genes, but unless those same people disapprove of all food that humans have moulded from wild ancestors – which means just about all the food we eat (organic, traditional or GM) – the view seems hypocritical. Other opposition may arise simply because some people enjoy pursuing a campaign – any campaign.

This is not to say that by comparison the pro-GM lobby looks virtuous. The global impression in this case is that the movement is driven more by basic commerce than by any true desire to help the world. Everything is marketing – nothing is altruism. And for a while in the 1990s there was a distinct element, particularly in Europe, of big businesses trying to inveigle GM foods into people's mouths without their knowing.

The interesting aspect for this book, though, is the way that both sides have tried to use inadequate science to argue their case. Of course, scientists did the research that initiated GM technology in the first place. And as problems arise, each is dealt with in turn to the best of the scientists' abilities. But tiny experiments in laboratories and studies of crops and wildlife in tiny fields are incapable of predicting or resolving any global issue. Current reassurances – just like current fears – are based on little more than guesswork, opinion and dogma. The only convincing reassurance – or otherwise – will finally come from the massive nationwide experiments on the people and environments of the United States and China.

8

Fragile Wills, Genetic Leftovers

People are responsible for their own actions. They have 'free will'. This simple hypothesis is the bedrock of human society the world over. Without it, there would be no logic behind the punitive element in legal systems – or, for that matter, the punitive element in many religions. Free will must exist; we are told so by our parents, our schools, and our laws – and by our own introspection. The thoughts in our head and our day-to-day decisions could not be due to anything else.

But is this so? Where is the scientific evidence that free will does – or does not – exist? And if free will were nothing but a dream, how would society cope with a rude awakening?

Genetic Appetite

Science is never more fragile than when there is no science – and there is no science of free will. We have philosophical meanderings. We have untestable statements of faith. But we don't have the proper, scientific study that we have demanded – albeit often in vain – for every other subject covered in this book. There are no controlled experiments; there are no brain scans. How could there be? How could we compare the behaviour or brain patterns of people with and without free will? We can't even experiment on other animals – because we are reluctant to endow them with the same dimension as ourselves.

So free will defies scientific study – but is nevertheless under threat from science. As genetic science becomes increasingly accomplished and shines its spotlight ever more brightly on human behaviour, it leaves fewer and fewer shadows in which free will can hide. The science of genetics is gobbling up the human persona – leaving us prodding for free will among its leftovers and panicking over what to do if it cannot be found.

Temperamental Genes

People differ, not only in the way they look but also in their temperament. Some are bright, others slow. Some are gloomy, others optimistic. Some are exciting to be with, others boring. Temperament can have many facets – and each facet influences the way its possessor behaves in this or that situation. On the whole we wouldn't expect a person with a low IQ to become a university professor, a meek person to become a fighter pilot, a shy girl to become a table-dancer, or a cubicle dweller to skydive. It does happen, of course, but when it does, it's a surprise – because the behaviour does not sit comfortably with the temperament. But if temperament determines behaviour, what determines temperament – genes or environment? Or can a person exercise their free will and take control of their whole approach to life?

Genetic studies are running away with the human brain. Bit by bit, the way we think and feel are being reduced to the genes we possess and the chemicals they produce. Each study, though, usually concludes that only part of the differences in temperament between people – usually about half – can be attributed to genes. For decades, studies of intelligence, as measured by the Intelligence Quotient (IQ), have been suggesting that around 50 per cent of the variation between people is due to differences in genes. And we have already seen, when discussing depression, that the temperament trait known as emotional sensitivity also has a genetic basis. Studies of twins showed that 50 to 60 per cent of the variation in this trait between children is due to the different genes they possess. A similar conclusion has also been reached for a trait known to psychologists as sensation – or novelty – seeking. This includes disparate facets that nevertheless seem linked. On questionnaires designed to assess temperament, high scorers for novelty seeking claim to find pleasure in varied, new and intense experiences. Such high scorers aren't necessarily fond of risk – but they will take risks for the reward of a new sensation. In contrast, low scorers prefer familiar, conventional and less intense experiences: they are cubicle dwellers. Studies of twins suggest that genes account for just over half (nearly 60 per cent) of person-to-person variation in novelty seeking.

To many people, such claims will seem bizarre – even far-fetched. How could genes influence something as abstract as temperament? The answer is simple: genes are responsible for brain chemistry and – as we saw for depression – brain chemistry dictates how a person feels and responds in different situations. So if people differ in their genes for brain chemistry, they will feel different about a wide range of things – including novelty. The brains of high scorers for novelty seeking must feel comfortable only at an excitement level that would make a low scorer feel anxious. The two brains must be different; the two brain chemistries must be different. And just as the key for unlocking variation between people in emotional sensitivity and depression turned out to be differences in serotonin levels, the key for unlocking variation in novelty seeking turns out to be differences in dopamine.

Dopamine is the so-called pleasure chemical: the brain releases it during sex, after a good meal, and in response to cocaine and amphetamines. It is a neurotransmitter – a monoamine – and it activates a variety of behaviour patterns. It motivates humans – and other mammals – to seek out things that feel good, and then triggers the sensation of pleasure when those things are found. All this takes place deep in the brain, in the region known as the nucleus accumbens. It is the release of dopamine here that gives the sensation of pleasure.

This influence of dopamine has been unravelled mostly via experiments on laboratory animals. For example, colonies of mice have been genetically modified to produce different levels of dopamine. In some colonies, the rush of dopamine was engineered to last 100 times longer than usual. These animals were in a perpetual state of hyperactive exploration, seeking novelty everywhere. Even in a very familiar cage, they behaved as if there was always something interesting just around the corner. They were the ultimate novelty seekers. Other colonies were engineered to lack the enzyme that allowed their brain to make its own dopamine. These mice were so lethargic they simply sat in the middle of their cage doing nothing. After two weeks, they could no longer be bothered to eat, drink or groom – and they would have starved. However, when injected with L-DOPA so that their genes could make dopamine again, they made an immediate recovery.

There have also been studies of humans. Some people lose the ability to manufacture dopamine in part of the brain. These are the sufferers from Parkinson's disease, caused by a degeneration of the dopamine-producing cells in a brain region known as the substantia niagra. The physical symptoms of the disease are reduced mobility and a trembling of the hands, but there is also a personality change. Sufferers appear serious, stoic or quiet; they score low on novelty seeking but normal on other personality traits.

Variation in novelty seeking thus seems to be partly due to genetic variation in the dopamine system, producing differences in the ease and speed with which neurons communicate pleasure to each other. One of these genes seems to have been identified. It is the gene involved in producing the 'D4' type of dopamine receptor – hence the gene is known as D4DR. Chemically, the D4DR gene is very variable. The length of the gene seems to be the important factor, just as it was for the serotonin transporter and Ob genes linked to depression. Different lengths of the D4DR gene have different abilities to bind dopamine: the longer the gene, the weaker the binding. In principle, therefore, this gene is very likely to influence how people feel – and in three separate studies (in Israel, the United States and Canada) this is exactly what was found. The longer a person's D4DR gene, the higher was his or her score for novelty seeking.

A simple 'quick-and-dirty' calculation – the sort one might carry out on the back of an envelope – has suggested that if D4DR is part of the genetic determination of novelty seeking, its variation explains about 10 per cent of total genetic variation in the trait. In other words, D4DR looks as though it might be one gene among about ten that all influence this facet of temperament. The search is on for the others.

Potential Sanctuary: Shared Environments, Unique Experiences

Step by step, genetic science is laying claim to human behaviour. Yesterday it was intelligence, today it is temperament, and tomorrow – who knows? Genes, neurotransmitters, receptors, neuronal activity and hormone production are being linked together to understand the human mental process. However, even the most aggressive evidence for genetic influence falls short of claiming to explain everything. It still leaves something of the human mind to be explained by other factors; and more often than not, it leaves a fair share. On the whole, around half of the differences between people apparently cannot be explained genetically. This would seem to be ample space in which to find an influence of other factors, such as the environment. It is also ample sanctuary for free will as it retreats from the genetic invasion. But before we can contemplate the latter, we need to quantify the former. Only then might we gain a better idea of how much of human mental activity and behaviour is left for free will to explain. So how much of the non-genetic variation can be proved to be environmental? To what extent do parents, schools, neighbourhoods and many other factors influence the facets of being human that we have considered so far?

We have to be wary of most direct studies of parental influence – because they invariably fail to allow for genes. As a result, they run a real risk of misinterpretation. We saw this for clinical depression: children raised by depressed parents are themselves likely to become depressed. Conclusion – a stressed childhood leads to depression as an adult. But the conclusion is wrong. As adoption studies have shown, parents pass on the genes for depression, not the example. We can only really trust studies that evaluate both genes and environment; ironically the best evidence for environmental influences comes from studies of genetic influence.

Usually, the environmental factors in genetic studies are divided into two types: shared and unique. Shared factors are those that influence both parents and children – or two children such as siblings or twins that grow up together. General parenting, socio-economic status, schooling and neighbourhood are all such factors. Their contribution to the differences between people is usually measured through adoption studies: comparing adopted children with biological and adoptive parents or comparing genetic siblings raised apart with siblings raised together. This last technique is particularly powerful when the siblings are identical twins.

In contrast, unique factors are those experienced by one child but not another; differences in the way siblings are treated by their parents fit into this category. So do the many events that can happen to children when they are apart. Unfortunately, unique factors cannot easily be measured in studies that also allow for genes and shared environments. What tends to happen is that researchers calculate the influence of the other two – then assume that all the variation that hasn't been explained (minus an unknown part due to experimental error) is due to unique personal experiences. This isn't ideal – but it is better than ignoring heritability and shared environments altogether.

The two types of environmental factor offer very different sanctuary for free will. In fact, shared factors leave no room at all. In just the same way that genes dictate characteristics irrespective of environment, shared environmental factors also dictate characteristics. If a person is exposed to a particular environment they develop a particular characteristic, come what may. That is, after all, how shared factors are identified and their influence measured. This means that the only part of human variation that could harbour free will is that due to 'unique' factors. Such variation is usually interpreted in terms of unique personal experiences but if free will has any role it will fit in here. Supporters of the free-will concept should therefore hope that all genetic and environmental studies leave a big chunk of being human to be explained by unique factors. So what do we find if we look closely at temperament? Which environmental factors are the more important – shared or unique?

By and large, unique factors are more important than shared. The latter have been shown to be a factor in determining a child's IQ score – which is good news for parents and teachers – but as we have already seen they have little, if any, influence on the predisposition to clinical depression. Nor do they have any detectable influence on emotional sensitivity or novelty seeking. This means that no matter how hard parents, schools and local councils try to create a particular home environment, school atmosphere or neighbourhood ambience, it will have little or no influence on whether a child grows up to seek thrills, live in a cubicle, feel shy or bold, or become prone to depression. So, with around 60 per cent of person-to-person variation in novelty seeking – and roughly the same for emotional sensitivity – being genetically determined, there is a whole 40 per cent (minus the obligatory experimental error) to be laid at the feet of unique factors.

In the case of emotional sensitivity, cumulative unique experiences seem slowly to erode the influence of genes: the contribution of unique factors increases from around 40 per cent during childhood to around 60 per cent during adulthood. Perhaps, for example, genetically shy people who have a series of positive experiences increase in confidence. The total change is relatively modest – and shows that even after the accumulation of life experiences, genes still play a significant part (40 per cent) in people's emotional sensitivity. Even so, there is a glimmer here that people might not be slaves to their genes for life. Something seems to reduce the influence of genes as a person ages. Of course, this something could simply be an irresistible influence of unique environment – but it could also be a manifestation of free will, flexing its muscles among the shadows of genes and environment.

This is dangerous ground, though. Here, the free will high wire is a precarious walkway. Those who take the route need care if they are to support their cause yet not seem unsympathetic to, say, depressives. To claim that a person could use free will to change their score for emotional sensitivity is tantamount to claiming that depressed people could snap out of their illness by 'getting a grip' on themselves. It could perhaps be argued that the depressed mind is ill and therefore robbed of free will whereas the mind of the high scorer for emotional sensitivity is not. In terms of brain chemistry and activity, though, the difference between the two is subtle, one turning into the other with ease.

Nature-Nurture: Conditional Result

The nature-nurture debate has raged for years. Now modern genetic science seems at last to have firm evidence that variation between people is due to both. Nature and nurture – genes and environment – are what produce human beings. Human attribute after human attribute is being identified as roughly half genetic and half environmental. Everybody should be happy. But the discussion hasn't ended. The nature-nurture debate has not produced a result. In fact, it has scarcely reached half time. The outcome may seem like a draw, but in fact nature is on a roll. It's a contest that nurture already cannot win. The genetic influence has been established, but despite the mirage, the environmental influence has not – not really. There are mischief-makers on the loose, posing as environmental effects. We can call them conditional genes.

Consider a pair of identical twins, Dick and Derek, with genes that score for emotional sensitivity somewhere in the middle of the spectrum. Their parents usually take them for walks in the park together, but occasionally, to encourage individuality, they take them separately. When Dick is two, he has his first close encounter with a dog – a friendly dog that looks cute, makes nice whining noises, rubs against his leg and chases sticks that his dad throws. The dog even lets Dick stroke it – and to Dick, it feels good. In contrast, the first dog that twin Derek meets is an ugly, vicious brute. It barks loudly, pushes him to the ground and even mauls his arm before dad can intervene. Dick grows up loving dogs – and other pets – and by the time he has his own family, his house is a virtual menagerie. Derek, though, shows fear from toddlerhood onwards. Whenever he sees a dog he freezes and hides behind whoever's nearby. Sensing his fear, dog after dog barks at him, chases him and a few get to bite him. Even as an adult, Derek is uncomfortable with dogs and never visits his brother's house.

This tale seems to be an open-and-shut case of the unique environment causing differences between people despite identical genes: a case of learning. But is it? Why did Dick respond positively to a pleasant experience and Derek negatively to a bad? The evolutionary logic is obvious. If something feels good, do it again; if it feels bad, avoid it in the future. It's a good rule, one that helped our ancestors to make the most of the countless different environments they had to cope with on their genealogical path to us. But where does that sensible rule reside? Certainly not in thin air – ultimately, it has to be genetic. There are genes – conditional genes – that say 'if this happens, change in this way, but if that happens, change in that way'. Dick and Derek, as identical twins, had exactly the same conditional genes – but because of their unique personal experiences, those same genes endowed Dick with one set of attitudes and behaviour patterns and Derek with another. It seems that identical genes can orchestrate even the differences between identical twins.

For the moment, conditional genes resist the conventional methods of investigation; their heritability cannot be estimated. The genes that show up in heritability studies are dictator genes – they force a trait on people irrespective of their environment and experiences. Conditional genes show up in twin studies as either shared or unique environmental effects. They could be sets of genes that can be switched on and off in different combinations by environmental triggers: maybe of four genes A, B, C and D, Dick's experiences had switched on A and B and Derek's had switched on C and D. Or they could be single genes that produce one effect when very active and another effect when less active, and the environment triggers changes in activity. It doesn't really matter how they work – as long as they do.

As an example, research on newborn rats has shown that altering the stimuli the youngsters receive can change neuronal development. If all the whiskers but one are shaved off, the neurons associated with that remaining whisker increase greatly in number, while those linked to the others atrophy. Similarly, if all whiskers are left intact but one whisker is repeatedly stimulated more than others, the neurons associated with that whisker again increase. It's a sensible response, one that moulds the rat's brain to receive future information in the best possible way. In later life, the rat will behave differently from one that received a different set of stimuli. But the difference between these two rats is not really due to the environment: if the rats hadn't possessed the relevant conditional genes, the environment would not have produced the change. The environment was no more than a trigger – the differences between the rats are entirely due to the instructions coded into their conditional genes. The same principle applies to Dick and Derek. If the twins had not possessed the conditional genes that said 'if attacked by dogs become phobic; if loved by dogs love them in return', their childhood experiences would have had no effect on their later behaviour. And if the conditional genes had been programmed with a different, less sensible, rule such as 'if attacked by dogs, grow green hair; if loved by dogs grow red', then that would have happened instead. The environment is continuously throwing up stimuli, but the ones that have an influence and the nature of that influence depends entirely on the instructions encoded in conditional genes.

Although identical twins like Dick and Derek will have identical conditional genes, doubtless most other people will vary – just as they vary in their dictator genes. So our search for free will's hiding place has struck a problem. Environmental factors may appear to overturn genetic influences – because they make people with identical genes behave differently and people with different genes behave similarly – but this power could be an illusion. Maybe, the environment – whether shared or unique – does no more than trigger yet further genes into action. The problem is that until we can find ways of studying the contribution of conditional genes to person-to-person differences, we cannot know whether only some – or all – of the differences between people are genetic. We are staring at a real possibility: that everything is in fact genetic. In which case, there would be nowhere left for free will to hide.

Incentives

How, then, can we continue our search for free will? So far, all we have done is go down two culs-de-sac. In chasing non-genetic differences between people, we have ended in a blind alley peopled by dictator and conditional genes. In chasing clear examples of free will in action, we have met only temperaments that don't really need free will. But maybe there is a way out of this latter dead end – a small path leading to a brighter place. Perhaps the problem is that there is not enough incentive for free will to resist traits like novelty seeking and emotional sensitivity; and perhaps free will itself does not have enough power to resist traits like depression. Maybe – once conditional genes have moderated, say, shyness or boldness to suit a person's life and lifestyle – he or she is content enough with their temperament. They feel no need to change it with free will. Perhaps, then, to see the true strength of self-control, we need to look at aspects of being human that have greater need of free will – aspects for which the incentives to combat genetic directives are much greater. As an example, we can look at people's battle against obesity.

Fat Dictators

Humans – like all animals – have always been preoccupied with food. If our ancestors hadn't eaten, if they hadn't consumed more calories than they burned, we wouldn't be here today. In past eras, it wasn't always easy to find and eat enough food – and it still isn't in parts of the Third World. Starvation was a real danger and, depending on our antecedents' environment, different genes evolved to cope with the problem. In places where food was scattered but ever-present, people were active, ate frequently, and burned off most of the calories they consumed by searching for the next small meal. In contrast, in places where food was briefly abundant but unpredictable, people gorged themselves when they could and stored the calories as fat to use slowly until they could eat again. Low metabolic rates and relatively slothful activities then helped them to conserve their calories while they waited.

Since these and all manner of other physiological and metabolic activities evolved, people have increasingly moved away from their ancestral environments. Global migrations have not so much mixed and matched people and environments but mixed and mismatched. Most seriously, many people with the genetic urge to gorge and conserve now find themselves in rich environments, such as the developed countries of the West. The genes haven't changed but the environment has. The average American, for example, now has a choice of 50,000 different food products compared with 500 in 1900 – and unlike in past millennia, all are localized in one easily accessible place, not scattered across the countryside. Many people don't even need to walk to collect their food. And easy food and the easy life are taking their inevitable toll on people's weight. According to the American National Centre for Health Statistics, one of every three adult

Americans over twenty-five years old is now obese – in other words they weigh more than 20 per cent above the ideal body weight for their height and frame. And the proportion is increasing – even as recently as 1980 it was only 25 per cent. A similar increase is occurring all over the Western world. No wonder, then, that obesity is increasingly a far bigger problem in some cultures than starvation – because obesity kills more people.

We have already seen, in discussing cholesterol, that obesity is linked to high blood pressure, coronary heart disease, some cancers and adult diabetes. The dangers are particularly severe for women: obese women are twice as likely to die prematurely than those of normal weight. Obesity and related conditions are the second leading 'preventable' cause of death after smoking. And people know it. Not only are they bombarded with health warnings over the dangers of becoming fat; they are also assailed with images of slim and desirable icons to show the rewards of staying thin. So obvious are the pros and cons to everybody that Americans, for example, spend $30 to 50 billion a year on weight-control (such as foods, books, and videos).

Here is a clear opportunity for free will to come out of hiding and show its face. The incentives are great – yet the evidence suggests that free will, if it exists, has a hard time over the matter. Studies of twins show that body mass index – a measure of weight relative to height – is 70 per cent heritable. Dictator genes, therefore, have a 70 per cent say in a person's weight. Even some of the finer details of obesity – such as the type and location of body fat – are inherited: for example, dictator genes account for 56 per cent of person-to-person variation in the amount of abdominal fat.

Conditional genes are also involved, but they seem uninfluenced by those aspects of the shared environment experienced as children, such as parental guidance and schools. This implies that parental efforts to instil good eating habits in their children are a waste of time. Studies of adoptees similarly find that people's weight follows their biological parents', not their adoptive. However, not all aspects of the shared environment are irrelevant.

Pima Indians now living in Arizona separated from their genetic kin now living in northern Mexico around 1,000 years ago. In Arizona, Pimas have the world's highest recorded prevalence of obesity whereas in Mexico, they are about average. Clearly, the availability of food and lifestyle can achieve through conditional genes what any free will alone cannot.

Some people have no need for self-control over eating. Their ancestral genes make them burn off excess calories no matter how much they eat. Others, though, are not so lucky. Their ancestral genes are forever urging them to gorge themselves and to conserve their energy so as to survive the hard times just around the corner that never come. In a rich environment, they desperately need to be able to overcome the power of their genes. Many try: around 70 per cent of American females aged fourteen to twenty-one years claim to be dieting. But to judge from the rising incidence of obesity, many fail. All over the Western world, the weight of people on diets yo-yos up and down as something within them struggles with their genes.

But what is that something? To the people concerned the contest inevitably feels like a war between will power and genes. It seems that the thoughts inside their heads must be their free will talking – urging them on as it tries to escape from its genetic prison. But despite how it feels, the reality could be quite different. The weight-watcher's internal struggle could be simply genetic indecision – the vacillation of conditional genes as environmental triggers switch them on and off. The person could be just an emotional observer, listening in to their genes' dialogue.

Those people who succeed in averting obesity feel that they have done so through the power of their will. There is, though, an alternative possibility: they may simply be different, genetically and by chance, from those who fail. Maybe their dictator genes are not pushing quite as strongly towards obesity as other people's. Or, perhaps, they have character traits of determination and directedness, traits that doubtless, like shyness and boldness, are genetic. Or they have just the right set of conditional genes for the situation. A set that can initiate a cascade of neurotransmitters and hormones capable of modifying their appetite and responses in the way they wish – as long as they encounter the right piece of information about risks, or the right range of rewarding images, or perhaps the right dietary regime. However a person feels as they struggle with their weight – whatever thoughts race around inside their head and seem to guide their actions – the dialogue could be no more than their genes trying to decide a proper course of action with the information to hand. For the moment, we cannot scientifically tell.

Violent End

Society won't crumble if science eventually squeezes free will from all roles in determining people's temperament. Even if personal struggles with weight – and perhaps related problems such as drugs – turn out to be solely genetic indecision, no social foundations will be irrevocably shaken. Environment will still have an input. Social changes and pressures can still fire this or that conditional gene. Society won't be impotent. Treatment and commercial enterprise could continue much as now. The most that would happen is that attitudes to such internal conflicts would change – and in general may even become more sympathetic. But there is one arena in which society is holding its breath, nervously waiting to see if science will do the decent thing and not rock the social boat. In matters of crime and violence, scientists are under great pressure to find a role for free will. How would the logic of punishment defend itself if science concluded that people really aren't responsible for their criminal actions? That any given person – if they have particular genes and particular experiences – cannot help but break the law. Yet the signs are already clear – in this respect as well, genetic science is gorging itself on what used to be the domain of free will.

Criminality is in a person's genes; so is being law-abiding. Studies of twins in North America, Germany, Denmark, Norway and Japan give an average heritability for adult criminality of 58 per cent. And there seem to be different genes for different types of crime. In Denmark, heritability was calculated as 76 per cent for property crimes and 50 per cent for violence against people. These are dictator genes: they push people into being criminal or law-abiding irrespective of environment. To judge from adoption studies in the United States and Sweden, though, most dictator genes work, at least during childhood, at the law-abiding end of the spectrum. The major such study in the United States compared the children of 'criminal' parents with the children of those who were law-abiding. The children were all separated from their genetic parents at birth or within a few days and adopted by non-related families. Those with law-abiding genetic parents were generally law-abiding themselves, no matter whether their adoptive parents were criminal or not. A few strayed from the straight and narrow, of course, but no more than in the population at large. With law-abiding genes, home environment made little difference: the genes are dictators during childhood. For the children of criminal parents, though, home environment made all the difference. In criminal homes, all measures of anti-social behaviour, including aggression, were several times greater. The genes were conditional: when in a criminal environment, become criminal; when in a law-abiding environment, become law-abiding.

The influence of conditional genes and law-abiding dictator genes declines with age; their function seems to be primarily to match children and adolescents to the environment they are forced to live in. When adult, after having had a chance to search for an environment appropriate to their dictator genes, these latter become the more important. It is then that dictator genes for criminality switch on and really begin to show themselves. This delayed appearance is not an unusual pattern for dictator genes. Many – the genes for baldness in men and menopause in women are examples – do not switch on until the person reaches a certain age. A study of thousands of twins who served in the Vietnam War showed that for juvenile crimes, dictator genes accounted for only 7 per cent of person-to-person differences and that the common environment, acting through conditional genes, accounted for 31 per cent, leaving unique factors, also perhaps acting through conditional genes, to account for the rest. For adult crime, 43 per cent was due to dictator genes and only 5 per cent due to the lasting influence of the shared environment. The influence of home, school and childhood neighbourhood quickly wanes once a person becomes independent. Instead, adults tend to switch to the level of criminality endowed by their dictator genes and their unique experiences.

So, how do the genes do it? How do they endow criminality or law-abidingness? One way seems to be by manipulating neuro-transmitters – at least as far as violence is concerned. Aggressive people have low levels of serotonin; so, too, do aggressive and impulsive mice, rats and monkeys. It has been calculated that about 25 per cent of monkey-to-monkey variation in aggression can be explained by differences in serotonin level. As we saw with depression, though, whereas dictator genes may endow people with different minimum levels of serotonin, conditional genes must also be at work, adding more or less to this minimum level in response to life events, such as falling in love. So, perhaps one of the ways the environment influences a person's criminality or law-abidingness is to trigger conditional genes to change the brain's serotonin level. Serotonin, though, can be only one of the links in the biochemical chain that leads to aggression – or the lack of it. There are others – hormones, for example – but ultimately, each one of them must begin with genes producing proteins of varying types and to varying degrees.

The search is on to identify these genes of violence. Some have been found, but so far they have been associated with only single-family lineages and are of little relevance to the population at large. The important genes are probably sprinkled throughout the human genome – but we know already that they are not evenly distributed. More than a fair share must be located on the Y-chromosome, the one that assigns gender – because the most important single factor in determining whether a person is likely to be violent or aggressive is whether they have a Y-chromosome. Women have no Y chromosome (they are designated XX); men have one (they are designated XY). Men are charged with five times as many aggravated assaults as women and ten times as many murders. Nor is this due to men being physically stronger than women – and therefore simply being aggressive towards the weaker sex. In all history, wars have been waged mainly by men against men. And if the frequency with which men kill men is compared with the frequency with which women kill women, the greater aggression of men is obvious. Whether we look at American cities, African villages, or Indian towns, men are over twenty times more violent than women. The ratio stays the same even when the background level of violence changes. For example, in England and Wales between 1977 and 1986, there were only around four killings per million inhabitants per year; in Detroit in the United States, there were 216 per million per year. Yet in both places, men killed men around twenty-five times as often as women killed women.

If there really are genes for aggression on the human Y-chromosome, then presumably before long they will be identified, along with more and more of the human genome. The best progress so far, though, has been made for mice: manipulating the genes on the Y-chromosome can dramatically change the level of aggression in males. There are no such neat or convincing demonstrations for humans, but there are strong hints. For example, the 1/1,000 men that are born with two Y-chromosomes – so that their sex chromosomes are XYY and hence they have a double-dose of Y genes – are up to five times more likely to end up in jail.

We know little of what the Y-chromosome genes do to heighten aggression in males – but we do know that they are responsible for the male's elevated levels of testosterone. Both men and women produce testosterone in their bodies. The genes involved – primarily, but of course not solely, on the Y-chromosome – are switched on in both sexes in the adrenal glands and the gonads (the testes in men and the ovaries in women). But because only men have the well-endowed Y-chromosome, they produce much more testosterone – the average healthy man has

260 to 1,000 nanograms of the stuff per decilitre of blood plasma; the average, healthy woman has 15 to 70.

Throughout mammals, testosterone is linked with heightened aggression. Castrate a male so that he produces only female levels of testosterone and he becomes docile; boost his testosterone with injections and he becomes aggressive. Most – though not all – experiments on men that change testosterone levels show the same association. Other studies show that naturally aggressive men tend to have naturally raised testosterone levels. In a study of American military veterans, those in the top 10 per cent for testosterone had records of greater antisocial behaviour, including assault and physical aggression. They also showed more drug and alcohol abuse – and had more sexual partners. The same association was found for hockey players and judo contenders.

The Y-chromosome genes involved in giving men more testosterone are largely dictator genes. They switch on in the foetus around the sixth week of gestation and cause a surge of testosterone to organize the formation of penis and testes. In adolescence, there is a second surge that may take testosterone levels briefly as high as 2,000 ng/dl, causing voices to deepen and hair to grow in adult places. They then stay on and give a man a characteristic 'resting' level of testosterone which declines with age. However, there are also conditional genes at work. The anticipation of sex, for example, raises testosterone levels. So, too, does competition – and winning keeps levels raised for a while even after the event. It doesn't matter what the contest, whether it's hard and physical like ice hockey, calm and cerebral like chess, or even trivial like tossing a coin. Testosterone aids competitive performance, and if the performance is successful it stays on hand to help again.

Testosterone doesn't explain all of male competitiveness and aggression. Pre-pubescent boys and girls, for example, have similar testosterone levels yet differences in aggression are already apparent. It is even possible that testosterone, like cholesterol, is more of a bystander than currently thought: one of the cast in the drama of aggression, but not the scriptwriter. The authors are the genes on the Y-chromosome, with a little help from others elsewhere. Nor can testosterone be the only actor – but it would need a determined critic to argue that the hormone played no part whatsoever. And, given the unequivocal link between testosterone levels and male libido, there can be little question that it also plays a part in one of the most specific aspects of male aggression – rape.

That rape is an act of aggression is axiomatic. It is defined, after all, as one person forcing their wishes on another in the face of unflagging resistance: usually, it is a man forcing sex on a resistant woman. And in some cases, rape may be no more than an act of aggression – an act that only incidentally uses sexual weaponry. But aggression per se explains only a small proportion of such violations. Rape has characteristics that demand additional explanation. In particular, the age distribution of women who are raped is very different from that of women who are the victims of non-sexual male violence. Specifically, men are most inclined to rape women at peak fertility – from their late teens to around thirty. They lust after this age-range in other ways too, of course – no matter what the age of the man – as illustrated by the pages of any girlie magazine, by pornographic videos, and by the wedding photos of geriatric but filthy-rich tycoons. Women in the target age-range are the most likely to conceive from a random sexual act, and evolution has shaped men genetically to direct their lust in this direction. The female victims of non-sexual male violence are much more evenly spread among age groups, though with a slight excess of older women.

There are undoubtedly dictator genes at work in rape, if only those on the Y-chromosome that determine maleness and sexual appetite in the first place. But there are also conditional genes. Rape involves males in cost/benefit reactions: they perform only when the potential benefits of rape seem to outweigh the potential costs. These are not purely cerebral calculations – because we can see the same facility in species with scarcely two neurons to rub together. It is also an emotional response, based on some inner appreciation of opportunity and risk. We see it in birds, we see it in apes, and we see it in men.

A man is most likely to rape a woman when the chances of success are great and the chances of recrimination – from the woman, her relatives or the wider society – are low. Hence, it rarely happens in broad daylight in a busy street; it usually happens when a woman goes alone with a man into a secluded bedroom, into a dark street or on a quiet country walk. Most rapes – date rapes – occur when a woman is alone with a man she knows; predatory rapes are less common and occur when she strays alone into the notice of a man alert for just such an opportunity. It happens when men are in gangs and rape is easy, and it happens particularly during warfare, when the gangs are armed and anonymous and the women are vulnerable.

In wartime, the increase in rape is because more men become rapists, not because a fixed number of rapists act more often. Evidently, more men encounter situations in which their personal threshold for rape is exceeded – their cost/benefit calculation comes out positive. One question – that has not yet been answered scientifically – is whether all men are capable of rape if an appropriate opportunity presents itself. Undoubtedly, men differ in their threshold to commit rape. Most men would not consider a lone woman walking down a filthy, icy back street to be a sexual opportunity – though a few do. At the other extreme, some men would not take advantage of a naked stranger, passed out through drink in a warm, dark bedroom – though many would. But could any man say honestly that there is absolutely no situation in which he might succumb and take advantage of a vulnerable woman? Many cultures are pragmatic over this. In the Hewa of New Guinea, for example, both men and women accept that if a man who is not closely related encounters a young woman in an isolated area, the chances are that he will rape her. There is no history of watching pornography or reading magazines to blame here. There is just a male, a young female, and an opportunity: one that the man's gene-driven urges push him into exploiting.

But is this really the sum of it: that the only difference between a rapist and a non-rapist – between a criminal and a virtuous citizen – is that the former have had their gene-created thresholds exceeded by random opportunity whereas the latter have not? Thresholds will differ – as everything genetic differs from person to person – and so will opportunity. So some – most – people will go through life without breaking local laws; others won't be so lucky. But if genetics and chance are the totality of the matter, miscreants have done no more to deserve punishment than the virtuous have to deserve reward. Neither had any control over their destiny.

If any of this is true – and if science ever convinces the society at large that it is true – the social repercussions take on nightmarish proportions. What would it do for women and the spectre of rape? How could we produce a fair and sympathetic system of constraint and punishment? How could the religions of the world adapt?

How Fragile?

But is this vision of humans as genetic automatons true? Are human beings nothing more than bundles of genes and environmental events? In the pages of this book, scientific study after scientific study – even those with enormous impacts on our lives – has been shown to be fragile. Why, then, should society even begin to consider giving up on the idea of free will simply because a branch of science apparently gives it nowhere to reside? Of course, it is legitimate for biological science to ask where free will could live in the human being if not in nerves, flesh and chemicals, the physical manifestations of genes. But, until the evidence that there is no hiding place is irrefutable, it would seem improper to talk of dismantling the world's legal systems.

So how robust are the genetic conclusions we have seen? Certainly, there are problems – many of them the same as bedevil the other studies we have discussed. First, the facets of humanity being studied are difficult to measure. Secondly, the associations between human behaviour on the one hand and brain scans and neurotransmitter systems on the other are mere correlations, with all the associated dangers of misinterpretation we have seen. Several spurious steps could exist between something as singular as a chemical or as complex as a brain signature and something as vague as behaviour. Finally, even the impressive associations between identifiable genes and behaviour are in fact no more than correlations.

These fragile threads, though, are bolstered by much more robust lines of research. Properly done, twin and adoption studies are difficult to interpret in any way other than a partitioning of genetic and environmental influences. On top of these, there are the genetic-engineering studies on other animals – such as on the dopamine genes and Y-chromosomes of mice. Whereas such studies may misjudge the range and nature of steps from gene activity to behavioural response, they are still more precise and powerful than most experimental techniques. Of course, studies on non-human animals are not always appropriate, as we saw for skin cancer and depression. It is difficult enough to diagnose clinical depression with certainty in a human, never mind diagnosing the same condition for study in a rat. Some of the behaviour involved in genetic studies, though, falls into a different category. Aggression probably does serve the same function in rodents and men. Sex certainly does – and so, in all probability, does rape. Tinkering with the genetics of behaviour in other animals probably is a reasonable way to study the genetics of human behaviour.

In its totality, the study of the influence of dictator genes on human behaviour, despite its flaws, seems almost as robust as biological science can manage for humans. It is certainly more robust than the non-existent science of free will. This doesn't prove that free will is fiction – but it does mean that supporters must do more than wave their hands and express belief. They need robust proof that genes are powerless – and, as yet, they have none. Of course, opponents of genetic science may justifiably question – as may we all – whether the heritability of human behaviour is really 50 per cent rather than 40 or 60 – but such nit-picking is trivial. The important point is that the influence of genes on human behaviour is neither zero nor total: dictator genes have some say in personal destinies.

Probably, then, the supporters of free will are wasting their time denying a role for dictator genes. The evidence in favour seems set to improve, not disappear. As the human genome is mapped and as more and more behavioural genes are located and identified, even the most die-hard opponents of genetic determinism will have to demur or look medieval. But there is no reason to give up – not yet. Those who believe – or just hope – that there is more than genes to being human would probably do better to focus their opposition on the evidence for conditional genes. The relationship between these genes on the one hand and environmental influences, learning processes, and long- and short-term memory on the other is still a tangled web with little actually proven. Just conceivably, free will could still be alive and well at the web's centre – tugging at the strings of human behaviour. And as no direct scientific test is possible, the hypothesis of free will cannot be disproved, only eroded. No matter how much we untangle the web, there will always be one small knot still intact that could be claimed to harbour self-control. Whether this small vestige is enough to justify whole legal processes is another matter – and beyond scientific remit.

Brittle Progress

This book seems to suggest that very often we can scarcely believe anything biological scientists tell us. Some will find this cheering because they can ignore gloomy predictions. Some will find it depressing because they cannot be reassured. The intention, though, was neither to cheer nor to depress but to illustrate simply that biology, as a science, is fragile.

Fragility has many causes. The most common arises from one of the favourite statistical tools for scientific investigation: correlation. All too easily, correlations masquerade as cause-and-effect. They are easy to do and even easier to interpret – and misinterpret. But understanding the biological reactions that produce a demonstrated relationship is much harder – sometimes impossible – and can take decades of argument, debate and experiment. In the meantime, despite the reverberant warnings, people so easily come to believe that they really have shown cause and effect. Almost every subject discussed in this book is in some sense fragile because all that supports the current understanding is a flimsy web of correlations. The most vulnerable examples are global warming and clinical depression, but the understandings of skin cancer and cholesterol are equally precarious.

Controlled experiments should be the most robust of scientific tools – but are sometimes no more than correlations in disguise. The role of statins in reducing heart attacks and cholesterol levels is a prime example. Another reason for controlled experiments to be suspect is when they involve extrapolation – particularly from one species to another. It is very difficult to experiment on humans: let's feed 1,000 people on British beef and see how many die! Maybe not – but often, therefore, there is no ethical alternative but to ask questions about human health via a panel of mice. Animal-rights supporters would not even consider this to be ethical. But just because something seems safe when tested on rodents doesn't mean it will inevitably be safe when given to humans. Other species often seem to have unique defences – like pot-bellied pigs to melanomas. Equally, just because something causes, say, cancer in rodents doesn't necessarily mean that it will do so in humans. Other species may have unique vulnerabilities – like, perhaps, rodents to UV light. Either way, it is dangerous to extrapolate conclusions to humans. Even controlled experiments on humans are not always what they seem, as illustrated by the treatments for clinical depression.

An even weaker route to scientific proof than correlation is the computer model. This is not to denigrate the role of computer models in the gradual build-up of scientific understanding – their use has provided valuable insights in many biological arenas. As a way of marshalling complex ideas and understanding interactions they are wonderful – but they do not prove anything. As anybody who has used such models knows, the desired answer is rarely further away than a tinkering with assumptions and formulae. Over the last two decades, the science of global warming has seen endless such tinkering, trying to make models fit data – yet still they produce only the crudest of approximations to the pattern of climate change during the twentieth century. As a result, they prove nothing about what might happen in the twenty-first.

Fragile foundations do not stop scientific subjects from growing. This is the house of cards effect: weak sources of evidence and weak ideas nevertheless bolster each other to produce a massive edifice. Usually, this happens because the whole construct seems to make some sort of intuitive sense – or it is in some way appealing, or has attracted commercial or media interest. The study of stomach ulcers was in this state before the discovery that the real cause was infection. Even now, there is a suspicion that the understanding of clinical depression, global warming, coronary heart disease and maybe even skin cancer is in the same state, just waiting for that breath of fresh air that will scatter the cards and start the building all over again on a firmer foundation – or not.

Not all of science's fragility, though, derives from the methods used. Biological truths may be elusive even when the search is robustly scientific – but when the search is victim to other factors, those truths can hide for decades. One of the worst mischief-makers is human impatience – particularly on the part of consumers eager to reap benefits. People want their lives to be easier, safer and healthier – and they want the improvement today, certainly no later than tomorrow. They find it frustrating – unacceptable – that science proceeds so slowly. When some apparently promising new treatment is not immediately put within reach of everyone, it is often regarded as being deliberately withheld. But hypotheses or products have to be not only conceived but also gestated. Then they are born vulnerable, more likely to die than thrive when challenged. Science needs time to explore each new possibility – and development simply cannot be hurried. Usually it takes twenty years or so for research to show the first glimmer of dependability – and if human health is involved, particularly if cancer is a factor, it can take much longer, perhaps half a century or more. Impatient consumers cannot wait that long.

Consumer impatience pressurizes governments and big businesses into hurrying science along, encouraging premature claims. Companies should not market new products quickly and claim that they are safe – because safety cannot quickly be proven. Governments should not demand robust answers overnight. During the British BSE crisis, it was totally unreasonable for the British government – and the British public – to expect the nation's scientists to know whether beef was safe to eat. Nearly twenty years on, we still don't know. The blunt truth is that no matter how many tests have been carried out on laboratory animals, until the general public has been exposed to something for a decade or so, the true level of safety cannot be judged. Consumer, commercial and government impatience conspire to force fragility upon science: scientists are pressured into a decision that is little better than guesswork. This is why thalidomide wreaked such havoc and why we can still do little more than cross our fingers while awaiting the consequences of sunscreens, British beef and GM foods. So far, there are no encouraging signs for sunscreen users – but at least there are no signs yet of Britons, Americans or Chinese dying in their hordes from what they have eaten. It is still too soon, though, to uncross any fingers.

It would be wrong to label only consumers, governments and big businesses with the sin of impatience. Scientists are not immune either. Like everybody, they would welcome fortune and fame. They are forever vigilant for the idea or discovery that will give them status – and when they get a contender, they don't really want to wait twenty to fifty years to reap the benefit. Deny it though many might, having an idea that is defendable and marketable in the short term is often more important than having one that proves correct in the long term. And, of course, science becomes fragile when it is more marketable than true. The persecution of cholesterol, which spawned a whole new food industry, and the study of skin cancer, which gave the sunscreen industry a new lease of life, may well be examples. So, too, may global warming and conservation, which have political market value and are being embraced by the developed world to pressurize the Third.

The converse situation can also lead to fragility: when the science is unappealing or easily discredited, even if it is true. Correct or not, every new idea is like a magnet, attracting the attention of that steel-hearted species of scientist, the demolition expert. For every scientist with an original idea, several more scientists will appear who dislike the idea – and often the scientist. Demolition can pay; there is sometimes more career mileage in destroying than promoting. The truth becomes secondary to career aspirations – and as in any contest, it isn't always the most virtuous who prospers. Of course, having people trying to find fault with something is a splendid way of testing its credentials. Opposition – justified or otherwise – is an important part of the scientific process, but it can sometimes backfire. Unscientific or unwarranted opposition can sometimes succeed in damaging a perfectly sound idea – as may well be the outcome for GM foods in Europe, at least in the short term.

Science's greatest fragility resides at its public interface. As scientists have such a difficult time discovering 'the truth' – and an even more difficult time believing 'the truth' as claimed by other scientists – it is not surprising that non-scientists take little notice of 'the truth'. The public listens to the most persuasive and appealing voice – not the most scientific – and even worse, they often listen only to people who say what they want to hear. Scientific argument is impotent outside of ivory towers: otherwise nobody would still claim the earth is flat, nobody would insist that there are gods, and nobody would reject the role of natural selection in shaping humans. At the public interface, perspective is more potent than science. How else could developed and Third Worlds show such different responses to conservation science? How else could people have such different views on free will? To some, murderers should be murdered and to hell with scientific talk of genes and environment; to others a murderer is the victim of their genes and environment and needs help and treatment, not punishment. The science is the same, the perspective different.

The media – and professional pressure groups – realized a long time ago that it is easier to tinker with people's perspectives and emotions than it is to educate them about scientific niceties. A good television programme or newspaper article that successfully taps people's strongest emotions will have far more impact than the most carefully argued scientific case. Even a scientist has to be impressed by the way the media squeeze more mileage from manipulating science this way or that than scientists ever have. The success in Europe of the campaign against GM foods is a prime example. Through emotive imagery (Frankenstein foods) and by belittling the benefits (tastier tomatoes), obscuring the risks (all tinkering with genes is dangerous) and throwing in some jingoism (blame American big business) a quasi-evangelical crusade was whipped into action. Maybe jingoism was the critical factor, because the same tricks haven't yet worked in America. The campaign over British beef or beef on the bone during the BSE crisis was equally impressive – and again, jingoism was probably the key to media success. The scientific evidence over risks was the same for all but the British public felt aggression towards their French neighbours for daring to call their beef dangerous and the French scorned the British for calling it safe.

The media – the shapers of perspectives – are the biggest cause of science's fragility. They can destroy good science and promote bad without any real conscience or comeback. After all, it may be twenty years or so before anybody knows whether the media's slant was justified or not – and they have a job to do: not only 'to inform the public' but also to sell copies or improve ratings. Just like many scientists, the media find the short-term marketability of science far more important than its truth and accuracy. If a science story doesn't interest, excite, scare, inflame, infuriate or entertain then it isn't a story. And if a slight distortion or a slight 'economy' makes a better story and sells more copies . . . But the media couldn't do any of these things without the complicity of scientists. Scientists view the media like other mortals view an errant but desirable spouse: they can't live with them and they can't live without them. Scientists coyly court the media, yearning for moments of exposure to the outside world but not wanting to appear forward. Then they complain bitterly when the media misquote, misrepresent or lampoon them, particularly if they misspell their name. But wait a week – or year – and they are back again, reluctantly agreeing to market their next idea.

Despite all of this – despite fragility and slowness, distortion and hype – biological science does eventually do the job required. Life is easier, safer and healthier than, say, a century ago – and although technology deserves much of the credit, biological research has also played its part. The worldwide increase in life expectancy is largely due to the real improvements that have been made in bio-medical science. Of course, there have been mistakes along the way: biological sciences have unintentionally killed and mistreated people in the past and will probably continue to do so in the future. Today's best hypothesis once properly tested may well be tomorrow's joke, even if it's not so funny. But eventually, the correct hypothesis will come along and stand the test of time, another robust pillar amidst the scientific colonnade of glass.

Further Reading

The following have been selected because they are available to the general reader (who may not have access to an academic library) and in most cases provide references to detailed academic papers for those, such as students, who may wish to investigate subjects in greater detail.

Sunscreens and Skin Cancer

Howard, W., *Attitudes to Sunbathing and the Risks of Skin Cancer*, London, Health Education Authority, 1997

Kenet, B. J., P. Lawler, et al., *Saving Your Skin: Prevention, Early Detection, and Treatment of Melanoma and Other Skin Cancers*, New York, Four Walls Eight Windows, 1998

Lane, W. I. and L. Comac, *The Skin Cancer Answer*, London, Avery Publishing Group, 1988

Long, W., *Coping with Melanoma and Other Skin Cancers*, New York, Rosen Pub Group, 1999

Cholesterol and Coronary Heart Disease

Julian, D. G., *Coronary Heart Disease: The Facts*, New York, Oxford University Press, 1991

Moyer, E., *Cholesterol*, London, Thorsons, 1998

Roth, E. M., *Good Cholesterol, Bad Cholesterol*, New York, Prima Publications, 1995

Zerden, S., *The Cholesterol Hoax*, London, Global Insights, 1997

Depression

Le Doux, J., *The Emotional Brain*, London, Weidenfeld & Nicolson, 1998

Pinker, S., *How the Mind Works*, London, Allen Lane, The Penguin Press, 1997

Wolpert, L., *Malignant Sadness: The Anatomy of Depression*, London, Faber & Faber, 1999

Wurtzel, E., *Prozac Nation*, London, Quartet Books, 1995

BSE and CJD

Dealler, S. F., *Lethal Legacy. BSE – The Search for the Truth*, London, Bloomsbury Press, 1996

Klitzman, R., *The Trembling Mountain: A Personal Account of Kuru, Cannibals, and Mad Cow Disease*, New York, Plenum, 1998

Rampton, S. and J. Stauber, *Mad Cow USA: Could the Nightmare Happen Here?*, New York, Common Courage Press, 1997

Ratzan, S. C. (ed.), *The Mad Cow Crisis*, New York University Press, 1998

Global Warming

Downing, T. et al., *Climate, Change and Risk*, London, Routledge, 1999

Houghton, J. T., *Global Warming*, London, Cambridge University Press, 1997

Rosenzweig, C. and D. Hillel, *Climate Change and the Global Harvest*, Oxford University Press, 1999

Conservation

Baskin, Y., *The Work of Nature: How the Diversity of Life Sustains Us*, Washington DC, SCOPE/Island Press, 1997

Burton, J. A. and G. Bertrand (eds.), *The Atlas of Endangered Species*, London, Apple Press, 2000

Eldredge, N., *Life in the Balance*, Princeton University Press, 1998

Kritcher, J., *A Neotropical Companion*, Princeton University Press, 1998

Mulner-Gulland, E. J. and R. Mace, *The Conservation of Biological Resources*, London, Blackwell, 1999

Paterson, M., *Global Warming and Global Politics*, London, Routledge, 1996

Spellerberg, I. F. (ed.), *Conservation Biology*, London, Longman Higher Education, 1996

GM Foods

Dibb, S. and T. Lobstein, *GM Free: Shopper's Guide to Genetically Modified Foods*, London, Virgin Publishing, 1999

Graham, V., *The EU and GM Foods*, London, Chandos Publishing Ltd, 2000

Lee, R., *How to Find Information: GM Foods*, London, British Library Publishing, 2000

Free Will and Genetics

Hamer, D. H. and P. Copeland, *Living with Our Genes: Why They Matter More Than You Think*, London, Macmillan, 1999

Thornhill, R. and C. T. Palmer, *A Natural History of Rape: Biological Bases of Sexual Coercion*, Cambridge, Massachusetts, The MIT Press, 2000

Index